T0054917

Glaciation: A Very Short Introduction

VERY SHORT INTRODUCTIONS are for anyone wanting a stimulating and accessible way into a new subject. They are written by experts, and have been translated into more than 45 different languages.

The series began in 1995, and now covers a wide variety of topics in every discipline. The VSI library currently contains over 550 volumes—a Very Short Introduction to everything from Psychology and Philosophy of Science to American History and Relativity—and continues to grow in every subject area.

Very Short Introductions available now:

Available soon:

For more information visit our website

www.oup.com/vsi/

David J. A. Evans

GLACIATION

A Very Short Introduction

OXFORD
UNIVERSITY PRESS

OXFORD
UNIVERSITY PRESS

Great Clarendon Street, Oxford, OX2 6DP,
United Kingdom

Oxford University Press is a department of the University of Oxford.
It furthers the University's objective of excellence in research, scholarship,
and education by publishing worldwide. Oxford is a registered trade mark of
Oxford University Press in the UK and in certain other countries

© David J. A. Evans 2018

The moral rights of the author have been asserted

First edition published in 2018

Impression: 1

All rights reserved. No part of this publication may be reproduced, stored in
a retrieval system, or transmitted, in any form or by any means, without the
prior permission in writing of Oxford University Press, or as expressly permitted
by law, by licence or under terms agreed with the appropriate reprographics
rights organization. Enquiries concerning reproduction outside the scope of the
above should be sent to the Rights Department, Oxford University Press, at the
address above

You must not circulate this work in any other form
and you must impose this same condition on any acquirer

Published in the United States of America by Oxford University Press
198 Madison Avenue, New York, NY 10016, United States of America

British Library Cataloguing in Publication Data

Data available

Library of Congress Control Number: 2018947005

ISBN 978-0-19-874585-3

Printed in Great Britain by
Ashford Colour Press Ltd, Gosport, Hampshire

Links to third party websites are provided by Oxford in good faith and
for information only. Oxford disclaims any responsibility for the materials
contained in any third party website referenced in this work.

Contents

Contents

Preface

Glaciers in their various forms are an important element of the Earth–atmosphere system, storing huge volumes of freshwater (10 per cent of the global total), interacting with, and potentially forcing, local to hemispheric-scale climate systems, controlling local to global sea level changes, and creating significant and lasting impacts on landscapes. The extent and intensity of each of these impacts varies through time, as glaciers and ice sheets grow and decay in response to climate-forcing mechanisms. In the present geological period of the Quaternary, often referred to as the Ice Age and covering the last 2.6 million years, glaciers and ice sheets expanded at times of maximum glaciation to cover up to 30 per cent of the Earth's surface. As a result, much of the developed land surfaces on which the large northern hemisphere human populations reside are composed of variable thicknesses of glacial deposits and glacially altered materials; it is therefore crucial to understand those deposits if we are to use them responsibly, for example as aggregate resources, groundwater reservoirs, and landfill sinks. Meanwhile, modern ice masses regularly remind us that we occupy and impact upon a climatically warming planet. Recent examples include the catastrophic break-up of the Larsen B Ice Shelf in Antarctica in 2002, the collapse of two mountain glaciers into monster ice avalanches in Tibet in 2016, increased iceberg production and ice recession around the margins of the Greenland Ice Sheet over the last two decades, and the disappearance of the life-giving mountain glacier 'water towers' of Himalayan agricultural communities.

Despite their significance, glaciers and the impacts of glaciation are not widely communicated at the more fundamental levels of education and general interest; a number of advanced texts are available on the state-of-the-art details on glacier physics, glacial geomorphology, and Quaternary glaciation, but the many significant recent advances in glacial research have not been compiled for, and targeted at, the introductory and general interest audience. Hence the need for this *Very Short Introduction*, which delivers a contemporary overview of the nature of glacial systems, their operation in space and time, the impacts of glacier ice on landscapes, and the main concepts that have developed in the study of glaciation as it relates to both longer (geological) and shorter (historical) timescales. In the context of the widely reported impacts of contemporary climate change on glacier health, this short introductory text will hopefully provide readers with an informed understanding of the operation of glacial systems in a changing world.

Since glacial theory was first established among earth scientists in the latter half of the 19th century the accumulating knowledge base has been compiled in textbooks that have placed glaciers and glaciation in a variety of contexts, including climate science, physics, and geology. This reflects the range of scientific disciplines in which glacier research has been undertaken, including geography, physics, geology, mathematics, and chemistry. Although this diversity of approaches and skill sets has made glacier science strong, it has also been inadvertently responsible for occasional faltering progress as a result of inevitable communication problems between scientific disciplines. Hence the occasional delivery of books that embrace the diversity are crucial to future cross-disciplinary communication.

While this *Very Short Introduction* will hopefully appeal to the informed lay reader who has become more aware of glaciers and ice sheets through intensive media coverage on climate change over the last few decades, it is aimed more specifically at students embarking on or in the early stages of university study of geography and related courses. A knowledge of glacial systems and glaciated landscapes is an asset in subject areas beyond their traditional home in geography; for example, in earth sciences, physics, engineering, and biological sciences. In these subjects in particular it is hoped that this *Very Short Introduction* will be an important narrative-based account of our knowledge of glaciers and glaciation. The final chapter on real-world applications of glacier knowledge and the future of glacier ice attempts to illustrate very briefly to a wide audience exactly why glaciers and their products are important as well as fascinating.

List of illustrations

Glaciation

List of illustrations

Chapter 1

Glacier ice: discovery and understanding

Glaciers: how and why?

Upon meeting the Antarctic Ice Sheet for the first time in 1842, Captain James Clark Ross aboard HMS *Erebus* named the imposing floating ice cliff in front of him the 'Great Barrier', as it appeared impossible to penetrate such a feature. Unbeknown to him this was but the outer floating shelf (later named the Ross Ice Shelf) of a huge ice sheet that flowed radially outwards from the intensely cold centre of the Antarctic continent. It was no doubt the occurrence of such a monolith of ice at the South Pole, in addition to the contemporary perceived state of permanent ice cover at the North Pole, that fuelled early proposals that isolated, huge 'erratic' boulders in northern Europe and Britain had been rafted there by vast armadas of drifting icebergs, breaking away from these higher-latitude ice masses during warmer periods and floating in radically deeper and more extensive oceans. How else could such boulders have been transported tens of kilometres from their outcrop source?

Access to a more plausible answer to this question actually lay in the glacier ice bodies that occupied and still occupy the high mountains of mid-latitude landscapes, features that were appreciated and understood long before the drifting iceberg theory emerged from the minds of hugely influential scientists such as Henry de la Beche (1796–1855) and Charles Lyell (1797–1875). By simply observing glaciers in terrains like the European Alps, explorers, scientists, and early geologists, like Horace-Bénédict de Saussure (1740–99), James Hutton (1726–97), Jean de Charpentier (1786–1855), and Louis Agassiz (1807–73), drew attention to the fact that the impacts (landforms) of these glaciers existed well beyond their present margins and hence they must have been more extensive in the past (i.e. they had grown in situ in these high mountain locations and did not require the frigid temperatures of the polar regions in order to exist).

Probably the most comprehensive study on the distribution of glaciers had been completed in 1795, the same year that Hutton published his influential treatise *Theory of the Earth*, by the underrated Icelandic glaciologist Sveinn Pálsson (1762–1840); but rather than be heralded as one of the founding fathers of glaciology, Pálsson remained relatively undiscovered until his book, *Draft of a Physical, Geographical and Historical Description of Icelandic Ice Mountains on the Basis of a Journey to the Most Prominent of Them in 1792–1794*, was translated from Danish into Icelandic in 1945 and then into English in 2004. Therein lay a clear appreciation that altitude, at whatever latitude, was critical to the inception and growth of glaciers.

Despite these early appreciations that glacier ice accumulated and flowed from mid-latitude mountains, it is a common misconception played out in many graphic reconstructions, even to this day for example in TV scientific documentaries, that glaciers grow in the highest and therefore coldest latitudes surrounding the poles and then march towards the mid-latitudes during phases of glaciation or ice ages. It is true that glacier ice builds up to considerable thicknesses at higher latitudes during phases of colder climate, but at the same time the environmental conditions of higher altitudes will spawn mountain icefields at much lower latitudes too; the occurrence of glacier ice on mountains such as Kilimanjaro near the Equator is a clear testament to this.

Glacier ice will form wherever the combined effects of precipitation (specifically snow), low temperatures, and topography create the optimum conditions. These optimum conditions are met in regions of high altitude and high latitude simply because they are the Earth's coldest places and generally therefore glaciers exist only at high altitudes in equatorial regions but occupy progressively lower altitudes towards the poles.

However, other conditions must also be satisfied, otherwise glacier ice would cover everything above these optimum latitudinal and altitudinal zones for ice growth. These conditions include aspect, relief, and the distance from the nearest moisture source. For example, aspect dictates that north-east-facing slopes in the northern hemisphere will receive the least amount of solar radiation and hence are more likely to be glacier covered first. Relief, or more precisely the steepness of terrain, is important because some mountain summits are too precipitous or narrow to hold thick snow and are instead blown free and hence are not glacier covered, even though their altitude puts them well above the threshold for glacier ice production. Moreover, if a region is located too far inland to receive a significant moisture supply it will fail to host glacier ice even though it is cold enough, simply because not enough snow falls.

Although our understanding of glacier ice inception and growth in relation to the combined effects of altitude and latitude is well established, a significant problem faced glacier scientists in the mid-20th century: that of ice sheet inception on continents. How and where did they grow in vast regions like the Canadian Shield, with its lack of substantial mountains, and how did they grow so fast to reach more southerly climes such as the Great Plains in North Dakota and Long Island, New York?

An extraordinarily prescient assessment of the likely dynamics of the former Laurentide Ice Sheet over North America was delivered in 1898 by Canadian survey geologist J. B. Tyrrell

(1858–1957), based upon years of painstaking field mapping of glacial landforms. He could explain the complexities of his mapping only by invoking multiple ice dispersal centres, which had grown in different regions of the Shield and then coalesced. Some sense of the physical achievements of this work can be gleaned from this extract from the *Barren Lands Digital Collection*, University of Toronto:

> In May 1893, J.B. Tyrrell set out for Lake Athabasca and the Barren Lands, accompanied by his brother, James, who acted as his assistant, and a party of six canoe men. After much difficulty and delay the party located and ascended the Dubawnt River north to its juncture with the Thelon River. They descended the Thelon River to Baker Lake and on to Hudson Bay. They then followed the coast south toward Churchill, but ice forced them ashore twenty miles from the settlement. Two of the party walked overland to Churchill and brought back a rescue party. After waiting in Churchill for the rivers to freeze inland, the party set out overland for Winnipeg which they reached on January 2, 1894. The Tyrrell brothers received much acclaim for their achievement and James Tyrrell wrote a book describing their adventures: *Across the sub-arctics of Canada*.

The lack of obvious mountain ranges to nourish such ice centres led the eminent Yale University geology professor, Richard Foster Flint (1902–76), in 1943 to prefer instead a single dispersal centre over Labrador in the east, with the ice sheet growing westward towards the predominant moisture-bearing winds. This became known as the 'highland origin/windward-growth' model. But were mountains essential in ice sheet inception? Research on ice and climate relationships in northern Canada in the 1970s suggested that they were not. This was encapsulated in a theory with the label 'instantaneous glacierization', proposed by University of Colorado glacial and climate researchers Jack D. Ives, John T. Andrews, Roger G. Barry, and Larry D. Williams.

The theory was born out of observations on an eight-year period in the early 1970s when late-lying snow over the Canadian Arctic effectively reduced regional air temperatures. It appeared that the high reflectivity (albedo) of the snow and increased precipitation had acted as a positive feedback loop, one that could be used to explain the formation of glaciers over any type of terrain but especially on low-altitude plateaux. The seeding points for an ice sheet were therefore seen to be numerous and the coalescence of multiple dispersal centres appeared to be the most feasible process of ice sheet inception; J. B. Tyrrell had been right all along!

A further model, one that is largely complementary to that of instantaneous glacierization, was presented by University of Maine glaciologist Terry J. Hughes in 1986 under the somewhat mischievous title of 'MITH' (Marine Ice Transgression Hypothesis). This simply developed our understanding that sea ice thickens over shallow marine embayments, a process that operates around the margins of the Antarctic Ice Sheet today. It involves the grounding of thickening sea ice to form first 'ice shelves' and then 'ice rises' and 'ice domes', which increase albedo and cause regional temperatures to drop, eventually bringing the snowline or *equilibrium line altitude* (ELA, the altitude at which a glacier can exist) down to sea level.

Balance and equilibrium

An understanding of how glaciers operate as systems was developed very early on in the history of glacier studies, with the work of de Saussure and Agassiz, as well as their contemporary Scottish physicist, alpinist, and pioneering glaciologist, James D. Forbes (1809–68). They made clear descriptions of what we now call 'glacier mass balance'. Professor of Natural Philosophy at Edinburgh University, Forbes undertook some of the first empirical studies on glaciers and is often overlooked as the pioneer of some glaciological ideas in favour of the more influential persona of Agassiz; their estrangement in 1841, over the premature publication by Forbes of observations made while visiting Agassiz on the glacier Unteraargletscher, did not help. A lack of substantial publications also hindered the development and acknowledgement of the influence of Forbes' work, the vision of which is aptly summarized by his, still largely valid, description of a glacier:

> A glacier is a mass of ice…which makes its way down to the lower valleys, where it gradually melts, and it terminates exactly where the melting…compensates for the bodily descent of the ice from the snow reservoirs of the higher mountains.

He continues to define the spatial arrangement of the various parts of the glacier surface:

> The snow line is a fact as definite on the surface of a glacier as on that of a mountain, only in the former case it occurs at a somewhat lower level. It cannot be too distinctly understood that the fresh snow annually disappears from the glacier proper. Where it ceases entirely to melt, it of course becomes incorporated with the glacier. We have therefore arrived at the region where the glacier forms; everywhere below it only wastes. This snowy region of the glacier is called in French 'névé', in German 'firn'. As we ascend the glacier it passes gradually from the state of ice to the state of snow.

All glaciers and ice sheets have a *mass balance*, a term which echoes those of economics, in which there is profit and loss and the balance between the two determines the health and dynamics of the system. Until the early 1960s the definitions of the various components of the glacier 'balance sheet' were only loosely formalized. This was effectively rectified at a symposium on glacier mass balance studies held at Cambridge University in 1962, where the most significant contribution on the subject came from the American glaciologist Mark F. Meier (1925–2012), who had been working on the mass balance of the South Cascade Glacier in Washington State. From there on a systematic and internationally recognized set of terms was quickly established. For any glacier the mass coming in to the system includes direct snowfall and indirect avalanches from surrounding slopes and is known collectively as *accumulation*. This mass is transferred downslope by glacier flow to locations where higher temperatures remove mass by melting and evaporation. This is augmented by iceberg calving into lakes or the sea. All of this mass lost to the system is collectively termed *ablation*.

The dominance of accumulation or ablation over a glacier surface defines two areas, known as the *accumulation zone* and *ablation zone* (Figure 1). These zones are separated by the *equilibrium line*, where accumulation equals ablation, the position of which will vary in altitude (ELA) according to the regional climate and local topography. Meier identified that the previously termed 'snow line' or 'firn line' (the line on the glacier above which the previous winter snow survived summer melting) did not necessarily correspond to the equilibrium line. This is because the altitude to which the snow melts each year can vary but the equilibrium line is somewhat more stable. A glacier will advance and its ELA will gradually lower when accumulation exceeds ablation (*positive mass balance*); conversely a glacier will recede and its ELA will rise when ablation exceeds accumulation (*negative mass balance*).

Another early perceptive view on glacier mass balance, and in particular its relationship to ice flow, was that of Harry F. Reid (1859–1944), latterly the Johns Hopkins University Professor of Dynamic Geology and Geography, who had been visiting the glaciers of Alaska and in the late 1800s to early 1900s published numerous papers on glacier fluctuations as well as seminal pieces entitled *The Flow of Glaciers* (1896), *Mechanics of Glaciers* (1896), and *The Reservoir Lag in Glacier Variations* (1905). Reid observed that there is a 'general law that the flow is less below, than at, the neve-line...this flow equals the product of the average velocity by the sectional area by the effective density'.

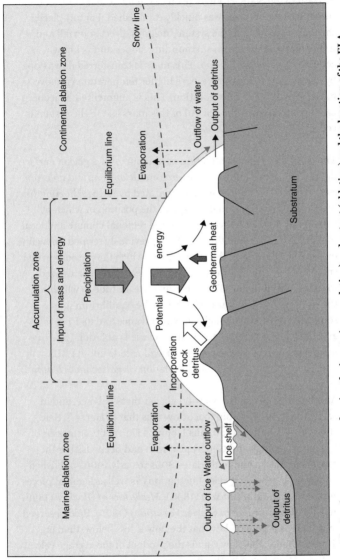

1. The ice sheet as a system, showing the inputs (accumulation) and output (ablation) and the location of the ELA.

It is now accepted that at any time during the life of a glacier it will be transferring mass from the accumulation zone in order to replace mass lost from the ablation zone, thereby initiating motion and maintaining the fastest flow and highest mass turnover at the equilibrium line. The rate of turnover of mass will vary according to the dominant glacier–climate interactions at the time and is encapsulated in the *balance velocity* concept. This concept works on the principle that a 'wedge' of mass lost in the ablation zone or snout of a glacier every year has to be replaced by a similarly sized 'wedge' transferred from the accumulation zone. Thereby glacier movement is required to balance the difference, and the speed of that movement is termed the balance velocity.

These principles of mass balance and balance velocity are critical to various processes of glacier motion as well as glacial debris transport and erosion as depicted in Figure 2. This illustrates the accumulation and ablation zones and the location of the ELA as well as the pattern of ice flow (grey arrows) related to the ice divide and local topography. Also important in this simplified diagram is the area of concentrated glacial erosion where an *overdeepening* is created, defined as a closed subglacial basin that under non-glacial conditions would form a lake. The diagram also depicts the pattern of delivery of debris to produce moraines and ice surface (*supraglacial*) debris in the ablation zone. It was not only H. F. Reid in North America but also Luigi de Marchi (1857–1936) in Italy and Sebastian Finsterwalder (1862–1951; also regarded as the grandfather of glacier photogrammetry and survey) in Germany who made early proposals that the transfer of mass through a glacier was intimately related to the accumulation and ablation.

At this juncture we probably need to define the term *moraine*, as it is applied to debris concentrations both on glaciers and around their margins. The term is of French origin and was initially introduced by de Saussure to refer to the landforms created at

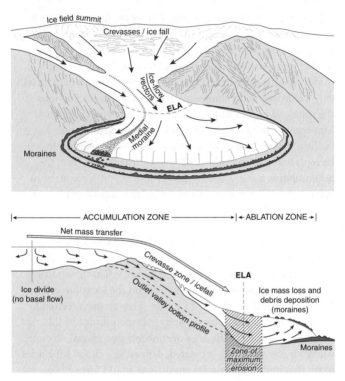

ACCUMULATION ZONE → | ← ABLATION ZONE →|

Net mass transfer

Crevasse zone / icefall

ELA

Ice divide
(no basal flow)

Outlet valley bottom profile

Ice mass loss and
debris deposition
(moraines)

Zone of
maximum
erosion

Moraines

Glaciation

2. **Simplified diagram to show a glacier snout fed by a mountain icefield.**

the margins of glaciers. He apparently adopted the term when he heard it used by Alpine farmers to refer to the features constructed along the margins of valley glaciers. It has since been widely used and misused, but glacial geomorphologists have gradually largely returned it to the place originally intended by de Saussure, especially because a moraine is a landform and hence must have been deposited by a glacier before it can be classified as one. Nevertheless, the term *medial moraine* (referring to a linear debris concentration formed between ice flow units in a single glacier; Figure 2) endures, despite the fact that it is used to refer to a supraglacial feature that has not yet been deposited.

The ice-making factory

In 1614 the English explorer and whaler Richard Fotherby pronounced with a perceptive eye and far ahead of his time that 'This huge ice is, in my opinion, nothing but snow, which…is only a little dissolved to moisture, whereby it becomes more compact.' Indeed, four centuries later we now understand that as snow accumulates year on year the deeper layers eventually turn to glacier ice. This involves the gradual reduction in volume of air-filled pores so that the bulk density increases. Stages in the transformation of snow to ice then take place, involving changes from fresh snow (with a density of 50–200 kg m^{-3}) to *firn* (snow that has survived one melt season and is 400–830 kg m^{-3}) and finally to glacier ice (830–910 kg m^{-3}). The change from firn to ice takes place when interconnected air passages are sealed off, separating the air pockets to produce bubbles (Figure 3). The density increases that take place from this point onwards simply involve bubble compression. This process results in the gradual build-up of horizontal stratification or *foliation*, visible in glaciers as density changes but also in the form of layers of fine debris that has blown on to the glacier surface during the process of *firnification*.

This primary ice-making process will also trap the ash fallout from volcanic eruptions and other atmospheric pollutants, and retain a proxy record of climate change in the form of changing proportions of oxygen isotopes, essentially a sample of ancient air. The value of this environmental change record was recognized as far back as the 1960s based upon the recovery of ice cores from the centres of the Greenland and Antarctic ice sheets and has since been increasingly widely utilized in understanding the patterns of climate change over the last several hundred thousand

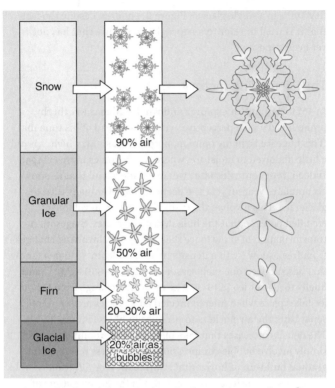

Snow

90% air

Granular
Ice

50% air

Firn

20–30% air

Glacial
Ice

20% air as
bubbles

3. **The production of glacier ice from fresh snow (50–200 kg m^{-3}) to firn (snow that has survived one melt season and is 400–830 kg m^{-3}) and finally to glacier ice (830–910 kg m^{-3}).**

years. However, such records are best recovered in the very few places where gradual ice deformation and flow are at their smallest, at the ice sheet centres. Predominantly such records do not last very long because of the inexorable drive by glacial systems to balance their budgets. Ice flow from accumulation to ablation zones brings about significant internal changes to the ice once it has been formed, especially where glaciers form over areas of high relief (Box 1).

Box 1 Ice cores: archives of climate change

In order to recover the most complete record of accumulation and indicators of climate change, such as those from the Greenland ice cores GRIP or GISP2 (covering the last glaciation and the last interglacial of 90,000 and 12,000 years duration respectively) and the Antarctic ice core EPICA (stretching back several glaciations and interglacials, up to 700,000 years), it is desirable to sample as close to the ice sheet centres as possible. This is because the ice sheet centres are divides from which ice will flow in opposite directions but be subject to only vertical motion directly below the divide. This has been illustrated by American glaciologist Richard B. Alley in his book on the Greenland ice core records, *The Two-Mile Time Machine*, in which he shows how the compression, stretching, and burying of ice layers can be predicted and then compensated in interpreting the stratigraphic record encapsulated in the ice sheet (see Figure 4).

Once a relatively undeformed and complete vertical stratigraphy of ice is recovered from the core barrel it is sampled at regular intervals for its concentrations of the two most abundant stable isotopes of oxygen, ^{16}O and ^{18}O, which account for 99.76 per cent and 0.2 per cent of the total oxygen in the Earth–atmosphere system.

Because ^{16}O is the lighter of the two isotopes being sampled, it can be evaporated from ocean surfaces more readily, so when falling as snow during glaciations it becomes trapped in relatively greater quantities in the growing ice sheets. Hence its concentrations relative to ^{18}O increase in those parts of the ice stratigraphy that accumulated during glaciations. So the variations through time can simply provide an indicator of periods of warmth and cold. Through this technique we know that the Ice Age or the Quaternary Period of the last 2.6 million years in duration has been composed of at least twenty glaciations of varying intensity.

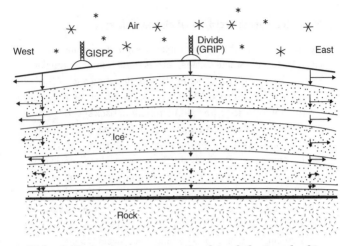

4. Richard Alley's summary cross section through the Greenland Ice Sheet to show the relationship between the GRIP and GISP2 ice cores and the relative motion of glacier ice as it accumulates over time. Note that ice layers thin towards the base of the ice sheet but the amount of thinning and stretching decreases in magnitude as the ice gets deeper and therefore older. The arrows show the relative vertical and horizontal motion in the ice.

Understanding the role of these internal changes in the creation of foliation or structure was accomplished on a range of glaciers in the 1950s and 1960s, for example in Iceland by British glaciologists Jack D. Ives and Cuchlaine A. M. King, on the Saskatchewan Glacier in Alberta by Mark F. Meier, and on the Pasterze Glacier in Austria by Norbert Untersteiner (1926–2012). From this period came probably the most influential overviews on glacier structure, published in 1960 by C. R. Allen, W. Barclay Kamb (1931–2011), Mark F. Meier, and Robert P. Sharp (1911–2004) and based on the Lower Blue Glacier in Washington State. They coined a vividly descriptive term, 'the structure mill', to convey the process of ice *breccia* production (crushing of material into angular fragments) at the base of ice falls, whereby a glacier becomes fractured and reconstituted before flowing away from the ice fall base.

So the foliation was not *sedimentary* (deposited in layers) but rather secondary and related to ice deformation. It appears commonly on most glaciers as a series of nested spoons, an analogy originally coined by James Forbes to describe the Rhône Glacier. The ideas of Forbes were being formulated at the same time, but often are not as well credited, as those of Agassiz, the widely acknowledged grandfather of the glacial theory. For Forbes, glacier structure made sense when you viewed ice as a *polycrystalline substance* (composed of myriad different-sized crystals) which underwent realignment of its crystal structure when deformed, in much the same way as *cleavage* (densely spaced, parallel parting structures) in metamorphic rocks.

We now understand that the structures and bands within glaciers are produced by a variety of mechanisms associated with their motion. Once in motion glacier ice is subject to localized velocity changes in response to alternating uphill or downhill sections of its bed. Bed steepening in a downflow direction subjects the ice to tensile (pull apart) stresses or *extension*. If the bed levels out or rises, however, ice flow slows down and the glacier is subject to *compression* (push together). As a glacier forming in a high-relief or mountain area descends from its upland catchment or accumulation zone it is forced to flow over many uneven surfaces, especially bedrock steps. The associated acceleration in glacier flow that is induced by ice descent into a topographic low point creates tensional forces, extension, and *crevasses* (Box 2). This is best demonstrated on alpine glaciers by spectacular ice falls, the most dangerous part of a glacier for mountaineers to traverse due to the extension, crevassing, and serac (ice spire) collapse.

Where ice flow once again slows and undergoes compression, crevasses close, and the very heavily fractured ice is reconstituted,

Box 2 Crevasses

The manifestations of tensional forces and extension imposed upon glacier ice as it flows across the variable topography are immediately visible, especially in mountainous terrain, in the form of crevasses. A crevasse is a deep fracture, generally up to 30 m deep in warmer or soft ice but potentially much deeper in colder and hence stiffer ice, especially if it is floating and creating icebergs. Crevasses are created by brittle failure in a rigid material like ice when it is subject to tension or simply pulled apart.

Various types of crevasse are recognized. *Chevron crevasses* are created where a lateral drag is created by valley walls on flowing ice. *Transverse crevasses* are a response to extending flow and they open up at right angles to the centreline of the glacier. *Splaying crevasses* are created in response to a combination of lateral drag and compressive flow. *Radial crevasses* develop in *piedmont lobes* (lobate glacier snouts that emerge from mountains and flow on to low terrain) due to lateral spreading and extension as well as longitudinal compression of the ice and are responsible for the characteristic re-entrants or indentations within the ice margin. These features are called *pecten*, after the ridged surfaces of scallop shells (Figure 5).

much like forcing together a crushed jigsaw puzzle, remoulding it, and producing a *melange* (ice and debris marble cake) and eventually a new foliation pattern. Hence the simple horizontally bedded structures produced by annual snowfall in the accumulation zone are completely modified and replaced with a new ice surface pattern called *ogives* or *Forbes bands*, named after James Forbes.

Ogives are arcuate, downflow convex bands or waves of alternating dark and light ice that are related to the annual formation of ice emerging from the crevassed zone of the ice fall, or the structure mill of Allen and colleagues. Since their intensive

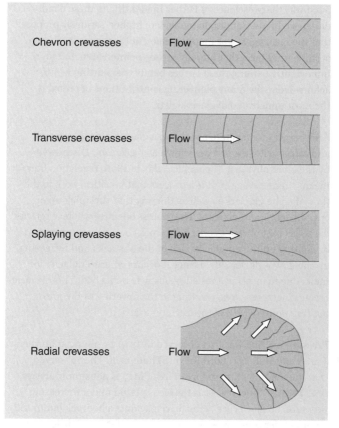

5. Crevasse types and their relationships to flow patterns in a glacier.

study in the 1950s and 1960s, including the acknowledgement by influential British glaciologist John F. Nye that they are part and parcel of the ice deformation process, a general consensus has emerged that the lighter bands relate to the winter, when snow falls into the crevasses, and the darker bands are created in summer, when more dust and other debris collects in the crevasses (more on the seminal contributions of Nye later).

Such was the prescience of Forbes in identifying these bands or ogives, Frank Cunningham, his biographer, seriously proposed that the term ogive was inappropriate, not just because it is an architectural term that literally means pointed arch, but more importantly naming them Forbes bands was a fitting way to acknowledge the many pioneering contributions of probably the most underrated of glaciologists.

Another way to generate new ice in a glacier is simply to melt and refreeze it via a process known as *regelation*. Discovered by the great physicist Michael Faraday in 1850, regelation literally means 'to freeze again'. It is a process that is critical both to the flow of some glaciers as well as the way that they pick up or entrain debris at their soles. The process operates wherever glacier ice slides past obstacles on a hard (rock) bed via a combination of melting on the up-glacier side of the obstacle, and refreezing on the down-glacier side. Hence it occurs wherever glacier ice melts due to meeting a resisting obstacle on its bed. This obstacle creates high pressures, resulting in the lowering of the *pressure melting point* (PMP) of the ice.

Pressure melting was first observed in 1849 by physicist brothers James and William (Lord Kelvin) Thompson. They observed that the melting temperature of ice (PMP) is not simply always 0°C, because it decreases as the ice is placed under increasing pressure. So glacier ice being driven against a bedrock bump will be put under pressure and may melt even if it is well below 0°C. The resulting meltwater then migrates to the low-pressure zone down-glacier of the obstacle where the ice is again subject to a higher PMP; here the meltwater refreezes to the glacier along with any debris picked up from the bed and completes a process that effectively amounts to ice bypassing obstacles by temporarily turning to water. Although debris is entrained in this way, this process cannot explain the very thick debris-rich basal ice sequences seen in many modern glacier snouts.

A far more effective way to create thick debris-rich basal ice sequences is by *glaciohydraulic supercooling*. Although it was discovered by Daniel G. Fahrenheit in 1724, supercooling remains a subject of significant debate in glaciology and has only relatively recently been identified as a significant process in glacier systems. It refers to the chilling of a liquid to temperatures below its freezing point without it turning into a solid. In simple terms this process operates wherever water flowing at the base of a glacier is forced to flow upslope and thereby undergoes a rapid pressure drop. This results in the water becoming supercooled relative to its surroundings and hence it freezes onto the glacier sole.

The theory of glaciohydraulic supercooling was first developed by Swiss glaciologist Hans Röthlisberger (1923–2009) in 1972 but it did not become well established as a prime mechanism for the production of ice/debris mixtures until it was applied to the remarkable stratified basal ice facies of the Matanuska Glacier, Alaska by a group of American glaciologists led by Richard B. Alley. Other ways of creating thick debris-rich basal ice include large-scale regelation in zones of abruptly changing glacier ice temperature (called *net adfreezing*; see the section 'How cold is a glacier?' later in the chapter) or glacier snout overriding of pre-existing debris–ice mixtures (called *apron overriding*).

These mechanisms of introducing debris to a glacier are critical to the effectiveness of the ice body in both eroding landscapes and depositing sediments. The earliest observations of debris introduction and transport include those of Agassiz in 1840 in his description of *lateral moraines* (moraines along glacier margins) and their convergence in valley glaciers to form medial moraines. Further views on how such moraines were not merely superficial forms but were instead linked to *englacial* (internal) structures and the bed were given by R. P. Sharp in 1948, who produced enduring diagrams to explain how such moraines simply marked the boundaries of ice flow units in composite glaciers.

The seminal study on the varied processes involved in medial moraine production and the catalyst for further assessments of the role of debris transport pathways and englacial structures in the production of debris-rich ice facies was that of N. Eyles and R. J. Rogerson in 1978. They proposed that, in addition to lateral moraine convergence (their ice stream interaction (ISI) type), medial moraines could accumulate by the melt-out of englacial debris *septa* (ablation dominant (AD) type), which are debris bands produced by material falling into crevasses and then being melted out further down the glacier, or material emerging from rock knobs being eroded by glacier flow; the term septa is used here simply to imply that the debris separates different ice flow units in the glacier.

Additionally they recognized that rock avalanches could dump debris onto the glacier surface (avalanche type, AT) and that continued ice flow would stretch out such accumulations. More recent assessments of the structures of glaciers have indicated that once debris is incorporated into a glacier, by whatever mechanism, it is subject to repeat folding and thrusting, so that its distribution over the glacier surface as it melts down is far more widespread than it would have been if no such ice deformation had occurred.

The patterns of debris accumulation as they relate to glacier flow are depicted schematically in Figure 6, which shows how debris is deposited in the glacier accumulation zone as horizontal layers and is then tightly folded by converging ice flow due to its movement into a valley constriction. This produces vertical debris septa and medial moraines originating by various processes that can be grouped under two broad categories. First, rock fall is entrained supraglacially and buried by snow or ingested in crevasses so that it appears as diffuse septum or clusters of rockfall debris. Second, debris is elevated from both the basal traction zone and the suspension or basal ice zone.

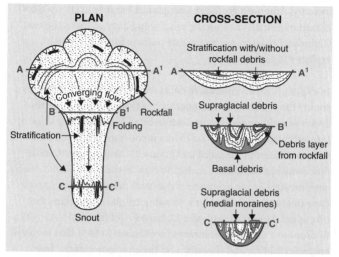

PLAN

CROSS-SECTION

Stratification with/without rockfall debris

A ————————— A¹   A ————————— A¹

Converging flow

B ——— B¹   Rockfall

Folding

Stratification

Supraglacial debris

B ——— B¹

Debris layer from rockfall

Basal debris

C ——— C¹

Supraglacial debris (medial moraines)

C ——— C¹

Snout

6. The typical patterns of debris concentrations in glaciers, showing how debris is deposited in the glacier accumulation zone as horizontal layers and is then tightly folded by converging ice flow due to its movement into a valley constriction.

This stacking of debris-rich ice is more prevalent in subpolar glaciers because they are not constantly losing basal ice to pressure melting and they are undergoing strong compression in their snouts. So in addition to ablation dominant type medial moraines, such glaciers also display a range of *controlled moraines* or debris ridges melting out directly from debris bands in their downwasting or lowering snouts. With the exception of supercooling, temperate glaciers are not effective basal debris entrainers and hence lack the thick sequences of apparently stratified debris-rich basal ice of subpolar and polythermal glaciers.

Such basal ice was the subject of much speculation by glaciologists involved in the earliest polar expeditions, one very important early example being that of the 1894 Peary Auxiliary Expedition, which included American geologist Thomas C. Chamberlin (1843–1928) as part of a small crew on the relief ship *Falcon*, sent to the east Greenland coast to secure the safe passage south of Arctic explorer Robert E. Peary.

In explaining the remarkable debris concentrations in the glacier snouts that he visited, Chamberlin was the first to propose the apron overriding mechanism, whereby an accumulating slope or *talus* of fallen ice blocks and debris is constantly overrun and reincorporated in the basal ice (Figure 7). Also apparent on the downwasting surfaces of such glaciers was the supraglacial debris zone created by the exposure of the basal debris-rich ice. It was the explanation of this debris on subpolar glacier margins that prompted American geologist Richard P. Goldthwait (1911–92) to propose a debris entrainment mechanism in 1951 that involved englacial folding and shearing and the emergence of the term *Thule-Baffin type moraine* for such a process-form relationship. An alternative process was proposed in 1961 by glaciologist Johannes Weertman, who pointed out that meltwater and debris would freeze-on at the junction of warm- and cold-based ice in the marginal zones of glaciers, a process that became known as net adfreezing. From that point onwards the complex thermal regimes of glaciers and ice sheets were recognized as being crucial not only to debris entrainment but also to subglacial mosaics comprising cold-based and warm-based ice.

How cold is a glacier?

The concept of a glacier being warm as well as cold was introduced in 1935 by the pioneering glaciologist Hans W. Ahlmann (1889–1974), who identified two principal types of glacier: those which are warm-based or temperate and those which are cold-based or polar. Between these two extremes he identified also high-polar glaciers with no surface melting and subpolar glaciers with some summer melting, thereby introducing the concept of polythermal or complex ice temperature conditions.

7. One of T. C. Chamberlin's photographs of debris-rich basal ice and apron overriding, taken at the margin of a Greenland glacier in 1894.

It may seem strange to refer to a glacier as 'warm', but glaciers vary as much in their temperature characteristics as the huge range of environments in which they exist. Indeed, individual glaciers will comprise a mosaic of areas of relatively cold and warm ice depending on their size and location. If a glacier is very large, for example an ice sheet, it will interrupt or trap the flux of geothermal heat that normally migrates from the Earth's core to the atmosphere, a concept first elucidated in the late 18th century by de Saussure as an explanation for glacier flow via sliding over its bed. This results in the melting of the ice that lies directly above the ground surface and so only the smallest and/or thinnest glaciers will not have water flowing at their beds due to melting.

Because of the varying influences of environmental conditions and glacier size on the temperature of glacier ice, glaciologists recognize a continuum of glacier thermal types including: (a) temperate (warm) glaciers, which are everywhere at the melting point except for a thin surface layer subject to seasonal freezing; (b) cold glaciers, which are everywhere below the melting point and are frozen to their beds; and (c) polythermal glaciers, which are composed of both cold and warm ice and are often referred to as cold or warm polythermal depending on the extremes of their climate locations. Weertman's 1961 model of net adfreezing was a clear example of the importance of polythermal glacier snouts to debris entrainment and transfer, the larger-scale implications of which were elucidated in 1972 by one of glaciology's most influential scientists, Geoffrey S. Boulton, in his benchmark paper 'The Role of the Thermal Regime in Glacial Sedimentation'.

The largest ice sheets actually melt so significantly at their bases that vast *subglacial lakes* can exist at their beds. The concept of large bodies of meltwater existing beneath ice sheets was proposed in the late 19th century by Russian philosopher Pyotr A. Kropotkin (1842–1921). Evidence for such lakes was collected in 1964 by a team of Russian scientists attempting to measure the depth of the ice using seismic soundings beneath their Vostok research station, but they missed the tell-tale signs of a vast subglacial lake beneath their feet and so it lay unrecognized and effectively undiscovered on the seismic printouts simply because glaciologists were not expecting to find one!

It was not until the 1970s that a multinational team led by Cambridge glaciologist Gordon deQ Robin finally identified subglacial lakes beneath Antarctica by using airborne radar techniques and christened the vast lake beneath the Vostok station 'Lake Vostok'. By the 1990s satellite altimetry, provided by the European remote sensing satellite ERS-1, had delivered more detailed information than ever on the subglacial lakes and the impressive size of Lake Vostok was confirmed, with an area of 14,000 km^2—twenty times the size of Lake Geneva. Eventually it became clear that more than 150 subglacial lakes exist beneath the Antarctic Ice Sheet, identifiable by areas of anomalously flat ice surface topography, and now the race is on to see if any of them support life in what would be the most inhospitable ecological niche on the planet.

Chapter 2
Glacier ice: definitions and dynamics

The glacier family

The shape and size of glacier ice bodies (their *morphology*) is dictated by the interplay between climate and topography, and a classification scheme has been adopted to cover all the variants that lie on a spatial and temporal continuum of morphologies (Table 1).

In any one location over time different glacier morphologies may evolve, thereby creating a temporal continuum, whereby glacierets and cirque glaciers form first on the coldest slopes and then become overwhelmed in turn by icefields and their valley glaciers and ultimately by ice sheets as glaciation proceeds. Even ice sheets evolve and never remain static systems, changing shape and thickness, undergoing surface slope changes and migrations in their ice divide locations and dispersal centres; such changes bring about abrupt shifts in the flow patterns of ice streams which have been recorded for us in the footprints of ancient ice sheets (see Chapter 7).

Glacier plumbing

Although glaciers are a manifestation of water in its solid form, water as a liquid is fundamental to glacial processes. The earliest indication that this was likely the case comes from the late

Table 1. The classification scheme for the continuum of glacier morphologies: the glacier family.

First-order classification	Second-order classification
Ice sheet and ice cap (unconstrained by topography or submerging the landscape)	*Ice dome* (multiple domes and ice dispersal centres are joined by ice divides)
Glaciers constrained by topography	*Ice stream* (relatively fast-flowing ice)
Ice shelves (floating ice masses fed either by outlet glaciers and/ or sea ice growth)	*Outlet glacier*
	Ice field (no dome and pierced by nunataks)
	Valley glacier (confined by a valley)
	Transection glacier (crossing underlying drainage divides)
	Cirque glacier (confined by a mountainside rock basin or cirque)
	Piedmont lobe (lobate ice body spreading into lowlands from the mountains)
	Niche glacier (small ice body on mountainside bench)
	Glacieret (tiny ice body occupying small depression or bench)
	Ice apron (ice body adhering to a mountainside)
	Ice fringe (ice body adhering to a steep coastal cliff)
	Confined ice shelf
	Unconfined ice shelf
	Ice rise (part of an ice shelf locally grounded due to its thickening)

18th century writings of de Saussure, who wrote: 'ice masses ... freed by the loss of contact with the bed on which they rest, freed by water from contact that ice could make with that same bed, sometimes even lifted by this water, must little-by-little, slide and descend following the valley slope'. In one of his typically far-sighted hypotheses, de Saussure implicated geothermal heat as the driving force behind the production of water at the glacier bed, assuming that the normal heat flux from the Earth's crust to the atmosphere would be impeded or partially trapped by the ice cover.

Although Agassiz in 1847 saw little merit in de Saussure's geothermal melting process, and reported the relative weakness of winter meltwater flow at the Unteraargletscher as evidence against such a process, it soon became apparent that even the thinnest and coldest glaciers may be characterized by the flow of water at their beds at some location, and all glaciers undergo melting at their surfaces when warmed in the summer or ablation season. The flow of this water through a glacier can be regarded as the glacier plumbing system and is driven by the melting of ice due to geothermal heat (subglacial melt), solar radiation (supraglacial melt), and frictional heat created by glacier flow (subglacial melt).

Glacier melting is generally very seasonal, with meltwater increasing in volume and discharge as it is augmented by snow and ice melt, rainfall, and water flow from surrounding river systems as the ablation season progresses. Additionally, in some circumstances large subglacial reservoirs or lakes can build up relatively slowly and drain catastrophically, especially where geothermal activity is concentrated beneath a large ice body. Glaciers may also interfere with the normal fluvial drainage of landscapes by damming valleys and forming ice-dammed lakes, which can also drain through the ice either gradually or catastrophically. Such catastrophic discharges, whether triggered by geothermal activity or lake overspills, are traditionally termed *jökulhlaups* after the Icelandic term to describe the many occurrences that are associated with the glaciers of Iceland.

The pathways of meltwater through a glacier change over the melt season, with drainage networks opening up and becoming more efficient as the melt season proceeds (Figure 8). Early-stage melting of snow and firn during the spring melt results in supraglacial run-off in short channels which plunge down through crevasses or *moulins* (vertical shafts). From this point the meltwater drains englacially, and as the melt season

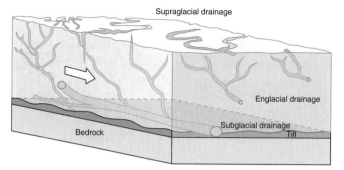

Supraglacial drainage

Englacial drainage

Subglacial drainage

Till

Bedrock

8. Idealized three-dimensional cross-section through a typical temperate glacier, showing the linkages between supraglacial, englacial, and subglacial drainage networks, which become increasingly better developed over the course of the melt season.

progresses the englacial drainage network develops through tunnel enlargements and the increased connectivity between moulins, tunnels, and crevasses. From here meltwater makes its way to subglacial tunnels or films, especially in temperate (warm) glaciers, where it plays a significant role in glacier flow because it reduces ice–bed friction or coupling.

The nature of this subglacial water flow and its influence over glacier dynamics varies according to the stage of the melt season but also in relation to the nature of the bed. A hard rock bed will tend to be associated with a thin water film with only occasional channels. As glaciologist Johannes Weertman demonstrated in the 1950s, an extensive water film will act to lubricate the ice–bed interface and to induce greater glacier sliding. Films are however unstable and so the water will usually organize itself into channels or even cavities, as highlighted by Weertman's contemporary, glaciologist Louis Lliboutry (1922–2007). On a soft glacier bed the water moves within the subglacial sediments, either as porewater at low flows or in shallow channels (*canals*) or even thin sheets at high flow (Figure 9).

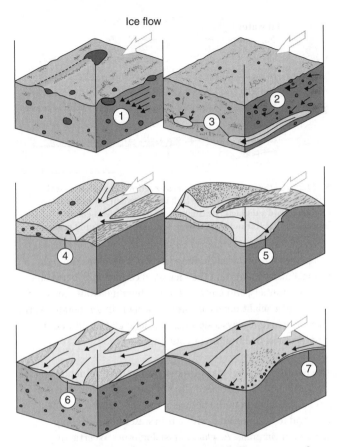

9. **The various styles of subglacial drainage: (1) bulk movement of water through deforming till; (2) Darcian porewater flow; (3) pipe flow; (4) dendritic channel network; (5) linked cavity system; (6) braided canal network; (7) Weertman film.**

The organization of meltwater into channels once it arrives at the glacier bed was analysed in considerable theoretical detail by glaciologists whose names are now inextricably linked with meltwater phenomena. In the early 1970s, the physics of subglacial water flow was essentially founded by the work of John F. Nye, Hans Röthlisberger (1923–2009), and Ronald L. Shreve. The identification by Nye of channels that become excavated into bedrock and thereby become fixed in their location at the glacier bed led to them being called 'Nye' or 'N channels'. Röthlisberger and Shreve recognized that some channels are excavated upwards into the ice (now known as 'Röthlisberger' or 'R channels') but, because the ice can creep and close them at lower water pressures, they are more mobile and often short lived.

Our bodies of knowledge on glacier hydrology and the dynamics of glacier flow are inextricably linked, because of the ubiquity of meltwater in glacial systems. So we now turn to the dynamics of glaciers and contemplate the implications of glacier hydrology for ice flow.

Creeping ice and sliding beds

Early observations by European Alpinists that glaciers appeared to be moving were put to the test by Franz Josef Hugi (1791–1855), Professor of Physics and Natural History at Solothurn University in Switzerland. Using repeat measurements on a boulder on the medial moraine of the Unteraargletscher, which moved more than 1300 m over the period 1827–36, he demonstrated that the glacier was flowing. Such was the scepticism for glacier dynamics at the time, his observations were initially dismissed as merely reflecting the sliding of the boulder over the motionless glacier surface.

For those who accepted the concept of glacier flow, such as Agassiz, the ruling theory of the time was that the motion was via 'dilatation', which was thought to entail the migration of meltwater into the glacier, where it refroze and the resulting expansion in ice volume initiated progressive down-glacier transport. Hence Agassiz predicted that flow should be fastest at the edges of the glacier and at depth, because this was where water input was at its highest.

While Agassiz set about testing his theory by measuring flow markers on Unteraargletscher in the summer of 1842, Forbes set up a similar experiment on the Mer de Glace and rapidly delivered his findings; the most significant was that ice flow was greatest along the centreline of the glacier, a direct contradiction to what Agassiz had predicted. Forbes then used his observations to expand upon the existing 'viscous flow theory', first proposed by the French cleric M. le Chanoine Rendu in 1841 in an underrated monograph entitled *Théorie des Glaciers de la Savoie*.

For some researchers, especially geologists who saw glacier deformation as a solid state structural problem, viscous flow was an unsatisfactory explanation for glacier movement, especially as glaciers displayed clear evidence for thrust faulting. For example, Rollin T. Chamberlin (1881–1948) of the University of Chicago's Geology Department embarked upon a field programme to demonstrate the role of thrust faulting in ice motion, a concept developed from observations made in Greenland in the late 19th century by his father Thomas C. Chamberlin. He employed a simple device to measure slippage along observed fault planes in glaciers and despite only moderate success then prematurely dismissed viscous flow.

To give credit where it is due, Forbes intimated that he thought ice behaved like a substance somewhere between a viscous fluid and a plastic solid, a proposal that proved to be remarkably close to our modern understanding of ice deformation. However, he has been traditionally lumped into the group of viscous flow proponents. Real progress on the nature of ice flow and rheology did not start until a benchmark meeting in London in 1948, reported in the first volume of the *Journal of Glaciology*. It was a joint gathering of the British Glaciological Society, the British Rheologists' Club, and the

Institute of Metals, and it was from the non-glaciological contributors, particularly Egon Orowan (1902–89), a materials physicist, that a law for ice flow was proposed; this was that polycrystalline solids, like ice, will deform plastically.

The 1948 London meeting was to have a career-defining impact upon one postgraduate student, John F. Nye, whose name has become synonymous with our modern principles of glacier flow mechanics. Nye reported on the findings of the meeting in the journal *Nature* and concluded:

> The discussion provided a happy illustration of the benefit of applying the simplest of physical reasoning to a science as yet largely empirical. In the words of one of the contributors, it was 'better to think exactly with simplified ideas than to reason inexactly with complex ones'.

For many this marked the entry of physicists into a research arena that had been traditionally almost the sole preserve of the alpinist, geologist, and geographer. Hence the notion of perfect plasticity of glacier ice was put to the mathematical test by Nye after the London meeting, leading to his 1951 paper entitled 'The Flow of Glaciers and Ice Sheets As a Problem in Plasticity'. This was the first of a series of papers by Nye and fellow countryman John W. Glen that defined our present understanding of glacier dynamics. What was needed after Nye's study was an experiment on ice deformation, and this was duly delivered by Glen in his 1952 paper entitled 'Experiments on the Deformation of Ice'. The laboratory results were surprising and revealed that neither Forbes' linear viscous nor Nye's perfect plastic behaviours were applicable, but rather, ice displayed a more complex relationship between stress and strain rate. Henceforth *ice creep* became the subject of the most fundamental of principles in glaciology; more specifically, this was *Glen's flow law*, named after its proponent John W. Glen.

The flow law explains the creep of ice in response to the stress imposed upon it, driven by the mass of the glacier and the gravitational force induced by its surface slope. For those who study glaciation and its myriad equations and formulae this is the one formula with which everyone has familiarity:

$$\dot{\varepsilon} = AT^n$$

where $\dot{\varepsilon}$ is the *strain rate* and T is the *basal shear stress*. It conveys the message that ice temperature (A) affects the strain rate so that ice deforms more readily the warmer it gets. It also shows that ice is a *non-linear viscous* material. That is, its strain rate or response to stress increases non-linearly with the applied stress, with a flow law exponent (n) close to 3. For John Nye, the derivation of Glen's flow law allowed him to propose 'The Mechanics of Glacier Flow' in the *Journal of Glaciology* in the same year (1952) and then to go on to test the new *flow law* on real glacier structures using ice tunnels and boreholes in 1953. In 1955 came Glen's paper 'The Creep of Polycrystalline Ice', followed by a sustained delivery of seminal papers by Nye on glacier motion, including the roles of ice falls, ice sheets, surges, mass budgets, and glacier hydrology.

Critical to the motion taking place at the ice–bed interface, the basal shear stress (T) is essentially created by a resistance to glacier flow or *basal drag*. Glacier flow is a response to the *driving stress*, which is controlled by the weight of the ice and gravity. This is best summarized by another equation that is readily familiar to students of glaciology:

$$T = \rho_i gh \sin \alpha_S$$

where T is basal shear stress, ρ_i is the density of ice, g is gravity, h is ice thickness, and α_S is surface slope. So this basal drag or shear stress, because it is resisting the driving stress induced by the glaciers' weight and gravity, is the cause of ice deformation. However, basal shear stresses vary considerably between and within glaciers in relation to the important variables of ice temperature, basal melt rates, and subglacial materials. For example, a deformable bed will help reduce basal shear stresses relative to hard beds and hence the overlying glacier snout profile may shallow or steepen respectively depending on how effective the bed is at resisting the stress imposed by the overlying ice.

Today we readily appreciate that the transfer of mass through a glacier from its accumulation to its ablation zone is driven by mass gained and mass lost in each of these zones respectively and this is facilitated by ice motion by not just one, but more commonly a combination of three principal modes: (1) *internal ice creep*, where ice crystals deform or slip past one another as a non-linear viscous material; (2) *basal sliding*, facilitated by basal meltwater production; and (3) *subglacial bed deformation* (Figure 10). The concept of glacier sliding has a long history but, in contrast, bed deformation has come very late to the glaciological research arena.

Basal sliding takes place when enough water is produced at the ice–bed interface to reduce the frictional resistance between

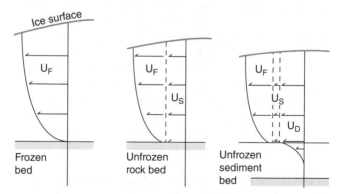

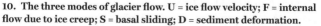

10. The three modes of glacier flow. U = ice flow velocity; **F** = internal flow due to ice creep; **S** = basal sliding; **D** = sediment deformation.

the two bodies. The concept was first proposed by de Saussure in his four-volume tome *Voyages dans les Alpes*, dating to the period 1779–96. From his observations on ice motion he posed the question 'can one doubt that this ice slides perpetually from high to low, and consequently has progressive movement', and hypothesized that geothermal heat flux was critical to the generation of water at a glacier bed.

Forbes' rejection of sliding in favour of his viscous flow theory did not deter the Cambridge mathematician William Hopkins (1793–1866), who set out to demonstrate sliding in the first physical experiments on such a process, culminating in papers on the subject in 1845 and 1862. Not only did he derive the first glacier sliding law but also related it to the process of regelation by recognizing that the melting point of ice decreases with pressure. This was to be verified by complementary field and laboratory experiments undertaken by geologists R. M. Deeley and P. H. Parr and published in 1914 in a paper with the somewhat unassuming title 'The Hintereis Glacier'.

It was not until the late 1950s to early 1960s that glaciologists Johannes Weertman and Louis Lliboutry were to further verify and elaborate upon the Deeley–Parr sliding law. But theory can only be put to the test by field observations and measurement, and one contribution stands out as the benchmark in this respect: that of Barclay Kamb and Ed La Chapelle in 1964, entitled 'Direct Observation of the Mechanism of Glacier Sliding Over Bedrock'. By excavating a tunnel to the bed of the Blue Glacier in Washington State, Kamb and La Chapelle directly observed the processes of ice creep and regelation around bedrock obstacles and measured the ice velocity at the bed in comparison to that in the basal ice and that of the glacier surface.

Stemming from the work of Hopkins, Deeley, Parr, Weertman, Lliboutry, Kamb, and La Chapelle, we now understand that glacier sliding largely takes place at small scales and only in certain locations, for example where pressure melting and regelation is operating around bedrock bumps or where water pressures are high enough to form films or cavities and act to locally decouple or separate the ice from its bed. The effectiveness of the latter process is immediately apparent in the spring events that are observed on temperate glaciers when their flow speed accelerates due to increased melt rates.

Galloping glaciers

Glacier behaviour that is particularly instructive for glaciologists is the dramatic tendency to *surge*. Surging glaciers are those which undergo alternate phases of rapid flow and slow flow or quiescence over cycles of a few years up to decades. Local inhabitants of areas containing surging glaciers were well aware of the unusually rapid advances of some glacier snouts before glaciologists had fully appreciated them. But what was causing this abnormal behaviour? The widespread nature of glacier surging was brought to the attention of the glaciological research community most effectively in the 1960s by the spectacular airborne glacier images of American mountaineer, glaciologist, and photographer Austin Post (1922–2012), in which were depicted the development of looped medial moraines created by the abnormally fast flow of individual ice flow units in complex valley glaciers.

A landmark in surging glacier research was the publication of twenty-six papers on surges from a very wide range of settings in a special issue of the periodical *Canadian Journal of Earth Sciences* in 1969, in which Austin Post and Mark Meier provided the first systematic definition of a glacier surge. Since that time numerous studies have been undertaken on surges, but the most instructive in terms of intensive monitoring have been on Alaska's Variegated Glacier. In particular a surge in 1982–3 was monitored by Barclay Kamb and a team of American glaciologists, who discovered that the surge was driven by the build-up of a subglacial water reservoir, which effectively decoupled the ice from its bed; the surge terminated when large volumes of the temporarily stored water were released rapidly at the glacier margin. This hydraulic model appears to be effective in temperate surging glaciers but a slightly different, thermal model might apply to polythermal surges, in which thickening ice in the accumulation zone may increase basal melting rates and cause the ice to surge forward over the ice of the frozen snout.

Deforming beds

Unlike basal sliding and internal creep, subglacial bed deformation is a relatively newly discovered process, having emerged only in the 1970s. Proof of its operation came from a daring experiment at the bed of the Icelandic glacier, Breiðamerkurjökull, by Geoffrey S. Boulton. It involved the excavation of a tunnel through the ice just above the bed so that four boreholes could be drilled downwards into the underlying glacial sediment, known as *till* (see Chapter 5). Strain markers (wooden pegs) were then implanted into the till at various depths and left for a period of 136 hours. When the pegs were then exposed in the walls of a trench created in the till, they displayed a pattern of downflow displacement, with the pegs towards the top of till having moved increasingly greater distances.

This proved that the ice and the till were coupled and that nearly 90 per cent of the forward momentum of the glacier was taking place via deformation of its bed, or what then became known as the *deforming till layer*. This deforming layer operates simply because only 40–45 per cent of till volume comprises mineral grains and the remaining space is occupied by water-filled pores; as the till is subject to significant changes in *porewater pressure* at the glacier base, even on a daily basis, its constituent grains can be forced apart or *dilated* by the porewater; in a shear zone like a glacier bed this causes the grains to effectively climb over one another.

After Boulton's Icelandic work it quickly became apparent during field experiments on modern ice streams in Antarctica that fast glacier flow was being greatly facilitated by till deformation. This was discovered during seismic investigations through Ice Stream B (later named the Whillans Ice Stream) by a team of American glaciologists in the 1980s, led by Richard B. Alley. The Antarctic ice streams, the fast-flowing arteries of ice sheets, have since become central to our understanding of both ice sheet operation and subglacial sedimentation and landform production, to the extent that the findings of Alley's team in 1986 were heralded by Boulton as a paradigm shift in the glaciology research arena.

What appeared on the seismic records from Ice Stream B was evidence of a layer of material separating the ice base from the bedrock. This could be explained only as a material with high porewater pressures and was therefore unfrozen and weak and susceptible to deformation by the overlying ice stream. Later attempts to retrieve a sample of this material via a deep borehole through the ice confirmed that it was unfrozen subglacial till. So the team concluded that ice stream velocity was maintained mostly by bed deformation. But a significant problem arose from this outcome—if the till was too weak to support the ice stream, why did ice flow not become unstable? The answer was provided in the 1990s by the introduction of the concept of *sticky spots*. Put very simply, the till did not cover the entire bed but instead thinned over bedrock bumps, which increased friction and acted as brakes on ice flow. Side drag at the edges of ice streams also slows them down.

In the 21st century, repeat satellite observations of our present-day ice sheets and improved and expanded radar, seismic, and borehole investigations into ice streams have uncovered more about their spatial and temporal variability and patterns of till movement at their bases. Probably the most exciting breakthrough for glacial geomorphology was reported in a series of papers over the period 2003–9 written by British Antarctic Survey scientists, who had measured *drumlins* (subglacially streamlined landforms) growing and migrating at the base of the Rutford Ice Stream. This discovery proved to be a game-changer for glacial geomorphologists who had long been theorizing on the evolution of subglacial landforms through the process of till deformation, or even the lack thereof, but had been frustrated by their inability to get beneath an ice sheet to test anything (see also Chapter 6).

Chapter 3
Glaciers through time

Ice ages, glacials, and interglacials

The term 'Ice Age' appears to have been the creation of German scientist Karl F. Schimper (1803–67), delivered in his 1837 paper entitled 'Uber die Eiszeit'. Shortly afterwards, in 1841, the German geologist de Charpentier published a map depicting a reconstruction of his 'glacier monstre' over the Alpine mountains and valleys of central Europe, which related to a former vast expansion of his contemporary mountain icefields. Together with the Swiss engineer Ignace Venetz (1788–1859), Swiss mountaineer Jean-Pierre Perraudin (1767–1858), and Danish geologist Jens Esmark (1763–1839), de Charpentier simply proposed that the evidence for more extensive former glaciers lay well beyond those of present ice masses in the form of moraines and erratics.

Such was the strength of scientific opinion at this time on the impact of *diluvial* events (the contemporary scientific nod to the notion of a biblical catastrophic flood), it was not until the later championing of the 'glacial theory' by Agassiz that the concept of multiple and more extensive glaciations was even partially accepted by the geological community. Indeed, even Agassiz himself initially had to be persuaded by de Charpentier and Venetz. It is unfortunate that Agassiz's charisma and celebrity status, as well as the publication of his extremely influential

Études sur les Glaciers merely months before de Charpentier's work in 1840, ensured his priority as the grandfather of the glacial theory over those who had instructed and convinced him of that theory.

Nevertheless, it is clear that as early as the first half of the 19th century, field observations on the extent of former glaciations had been well established. Formal recognition of multiple glaciations came in 1909 with the *alpine model* of Penck and Brückner (Box 3), who fully acknowledged that

Box 3 Multiple glaciations and the alpine glacial sequence

The earliest documentation of multiple glacial sediments separated by interglacial organic deposits was presented by James Geikie (1839–1915) in his 1874 benchmark work *The Great Ice Age*. By the time the third edition of his book was published in 1894 he had established that six glacials had taken place.

The influence of Geikie's work on Albrecht Penck (1858–1945) and Eduard Brückner (1862–1927) when they compiled their alpine glacial sequence from 1901 to 1909 is clearly evident in their dedication of the work to Geikie (Figure 11). The *alpine model*, as it became known, worked on the principle that the former maximum position of a stationary glacier is demarcated by a *glacial series* comprising the three elements of glacial basin, terminal moraine, and outwash plain. As the outer stretches of the outwash plains lie beyond the direct impacts of glacial erosion their occurrence in four stacked terrace levels ('deckenschotter') constituted evidence of multiple glaciations, which Penck and Brückner named after the local rivers: Günz (oldest), Mindel, Riss, and Würm. It was to prove so influential that for the next half a century of research Quaternary geologists believed that there had only ever been four glaciations.

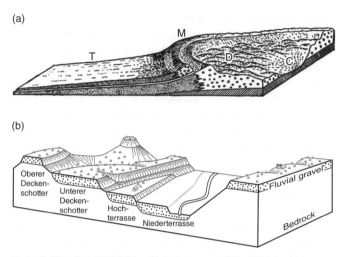

(a)

(b)

11. a: the Penck and Brückner (1909) concept of the glacial series, captured in an idealized sketch by Penck (1911). Marginal moraines (M), of which there are two, feed into an outwash plain (T) and are inset with drumlins (D) and an eroded basin (C). The superimposition of the moraines and outwash in this diagram is actually incorrect, because the outer moraine-outwash pair are older and hence should lie underneath the inner pair.

b: the appearance of multiple glacial outwash terraces or deckenschotter, each one linked to distant ice margins of different age.

some geologists had already identified evidence for cyclic changes in climate in glacial sediment sequences.

At the time of the development of the alpine model of multiple glaciations little was known about the possible climate drivers for such complexity in the glaciation record. Indeed two opposing schools of thought developed among those who had finally accepted the glacial theory; on one side stood the *monoglacialists*, who could accept glaciation had taken place but only in one complex event, and on the other side were the *polyglacialists*, who championed the cause for multiple glacial events.

It was clear even to monoglacialists that more than one glacier oscillation had taken place and that multiple glacial deposits (tills) could be observed, often associated with intervening soils or organics; but they employed a wide range of arguments to refute multiple glacials and interglacials, proposing instead that glacier margins had merely advanced over adjacent vegetation. The monoglacialists appealed to the *topographic hypothesis* of glaciation, an idea introduced in the mid-19th century and relating the development of glaciation to gradually increasing areas of high elevation terrain due to ongoing land uplift.

Since the erection of the alpine model, more than a century of research on the Earth's climate record (*palaeoclimate*) has established a firm understanding of the nature and pace of climate change over the last 2.6 Ma (million years) and the impact these changes have had on glacier development and extent. Figure 12 shows the oxygen isotope record, as extracted from very long sequences of ocean floor sediments, for the last 2.6 Ma, or the Quaternary Period of the geological column, the period we also more generally refer to as the Ice Age, even though much older ice ages or *Snowball Earth* states are evident in the geological record (see Box 4). This period has been identified based upon its large

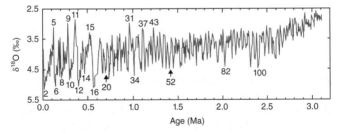

12. **The deep sea oxygen isotope record with selected marine isotope stages labelled. Note the increased importance of the 100,000-year cycle in the later part of the record after the mid-Pleistocene Transition at around 0.75 million years ago. Glacials are given even numbers and interglacials odd numbers.**

Box 4 Ice Age extents and Snowball Earth

Although the Quaternary Period is more commonly known as the 'Ice Age' (Figure 13), severe glacial conditions have dominated the Earth's climate during much older geological periods. Indeed, Earth's climate has been subject to contrasting conditions of *icehouse* (high average glacier ice cover) and *greenhouse* (little or no ice cover). We presently inhabit an Earth in an icehouse state, one that was initiated some 34 million years ago. Previous major glaciations occurred during the Permo-Carboniferous (326–267 million years ago), the Late Devonian to early Carboniferous (361–349 million years ago), and the late Ordovician (445.6–443.7 million years ago), with the most severe icehouses occurring during the Proterozoic Makganyene glaciation (2.3–2.2 billion years ago) and the glaciations of the Cryogenian Period (720–635 million years ago).

So severe were the icehouse conditions in the Cryogenian that glacier ice extended into equatorial latitudes and average global temperatures were well below freezing, creating what has become known as a Snowball Earth state. The control on such conditions is thought to be extreme CO_2 drawdown and the maintenance of subfreezing temperatures by large losses of shortwave radiation from highly reflective ice surfaces (i.e. an unusually intense ice–albedo feedback effect) (Figure 13).

volumes of global ice relative to previous geological periods and is the culmination of a long-term cooling trend that began some 55 Ma ago. A number of significant changes took place to initiate this plunging of the Earth's climate into the deep freeze including the building of the Himalayas, a long-term reduction in global CO_2, the formation of the Isthmus of Panama to shut off the Atlantic–Pacific connection, and an overall increase in global albedo or reflectivity due to increased snow and sea ice coverage.

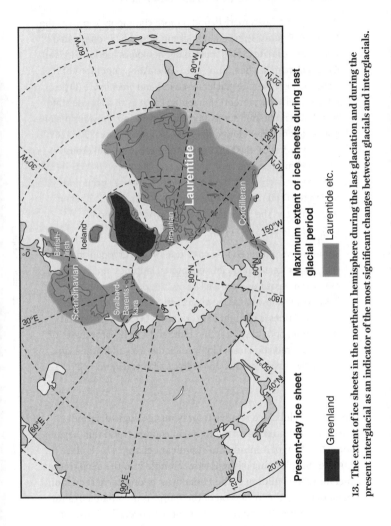

Present-day ice sheet

Maximum extent of ice sheets during last glacial period

Greenland

Laurentide etc.

13. The extent of ice sheets in the northern hemisphere during the last glaciation and during the present interglacial as an indicator of the most significant changes between glacials and interglacials.

46

By around 2.6 Ma ago the long-term cooling had settled down to the oscillating but relatively cold temperatures of the Quaternary Period. In response, the volume of global ice has oscillated during the Quaternary Period, as recorded by the oxygen isotope trace in Figure 12. This is measured using changes in the relative concentration of the heavier isotope of oxygen ($\delta^{18}O$) over time, as recorded in the skeletal remains of marine organisms that accumulate with sediment on the sea floor. This varies through time in comparison to the lighter ^{16}O, which becomes preferentially trapped in the constantly fluctuating volumes of global glacier ice and so is not always present in the same concentrations in sea water (see also Box 1 and the complementary record in ice cores). This isotope trace is used to identify phases of relatively high and low global ice volume, each of which are numbered as *marine isotope stages* (MIS 1–104 in the last 2.6 Ma), with the even numbers representing cold (glacials) and the odd numbers representing warm (interglacials).

So the Quaternary Period is characterized by two dominant climate extremes of glacials and interglacials but climate is never stable, oscillating instead dynamically from one extreme state to the other, prompting the use of the terms *stadial* and *interstadial* to refer to phases of relative cold and warmth during glacial stages. The intensity of glacials has also changed during the Quaternary Period with the dominant length of climate cycles changing from 40,000 years to 100,000 years at around 750,000 years ago. This has been termed the *Mid-Pleistocene Transition* and appears to have been responsible for the first appearances of the major northern hemisphere ice sheets, which require longer cold climate phases.

Upon reflection it now seems that the monoglacialists of the early 20th century were engaged in a perverse type of special pleading in attempting to explain away the land-based stratigraphies of multiple tills with intervening organic deposits and soils. Such evidence clearly recorded significant changes in environmental conditions and depositional processes, but in fairness to the monoglacialists the stratigraphies were, and still are, fragmentary and incomplete due to the erosional capacity of glacial processes and the occupation of approximately the same areas by repeat glaciations. Additionally, the monoglacialists did not have the luxury of being able to view the uninterrupted Ice Age story through the ocean core oxygen isotope records which began to emerge from ocean floor drilling programmes in the 1950s.

Glacials or glaciations are often viewed in terms of the maximum extent of ice sheets and glaciers when the Earth's glacier cover increases from 10 per cent to 30 per cent. But these maximum positions of ice margins are never achieved at the same time everywhere (they are time transgressive) and represent only very short periods of the cold marine isotope stages, possibly only a few hundred years in length. Additionally, ice coverage in reality actually pulses over short timescales, gradually building up to its very short-lived maximum extent so that glaciation as an earth surface process is concentrated temporally mostly on landscapes that are prime locations for more stable ice cover, such as upland areas. This is known as the concept of 'average glacial conditions' and simply reflects the fact that most of the Quaternary Period has been characterized by an earth surface with an intermediate style of glacier coverage, or a state somewhere between full glacial maxima and interglacial minima.

Importantly these intermediate conditions are the most effective in terms of longer-term landscape change because they dominate for most of the time and so equate to far longer periods of glacier occupancy than those associated with maximum ice sheet coverage. As an example, upland landscapes, because they are prime sites for early stages of glacier occupancy, are likely to have been glacier covered or glacierized for longer. Indeed many high mountains host ice during interglacials, as exemplified by the European Alps, Alaska, Andes, Himalaya, Rocky Mountains, and New Zealand Alps today, and hence it is unsurprising that the most impressive and mature glacial erosional forms occur in such alpine settings.

In this book it is the evidence of the most recent glacial events that is employed to demonstrate the principles and products of glaciation, because this evidence is the most fresh and less modified. Hence we will focus on the *Last Glacial Maximum* (LGM), which is the glaciation associated with MIS 2 and dating to 19,000–23,000 years ago, as well as the post-LGM ice oscillations collectively known as the *Neoglacial* or *Neoglaciation*.

Persistent versus temporary ice sheets

The constant oscillations in global temperature throughout the Quaternary Period have brought about significant changes in the nature and extent of the Earth's glacier ice cover, so that even the largest ice sheets can disappear or revert to mere remnants of their former mass. For example, the Laurentide Ice Sheet over North America builds up over the course of a glacial cycle through the coalescence of multiple dispersal centres, the last remnant being the present-day Barnes Ice Cap on Baffin Island. Similarly, the formerly vast mass of the combined Fennoscandinavian, Barents Shelf, and British-Irish ice sheets largely disappeared over north-west Europe and the European Arctic, leaving only remnants over the most northerly upland landscapes of Scandinavia, Svalbard, and the Russian Arctic islands.

Our largest present-day ice sheets of Greenland and Antarctica seem to be more permanent fixtures of the global *cryosphere* (that element of the Earth–atmosphere system that includes semi-permanent ice), and certainly the Antarctic Ice Sheet has persisted since it became a stable system around 10 Ma ago. However, Greenland may not be as persistent as its present interglacial existence might suggest. Recent evidence gleaned from the base of ice cores through the ice sheet indicates that it was reduced to approximately half its present size at the height of the last interglacial and likely disappeared altogether during the longest of our most recent interglacials (MIS 11) around 400,000 years ago. Present-day indications of accelerated recession of its margins suggest that it might be well on its way to that reduced status once again, and given present predictions for future climate change might even disappear again.

Fluctuating mountain icefields

Because of their altitude, mountain landscapes can develop glacier ice bodies irrespective of their latitude and indeed many have likely hosted permanent ice throughout most of the Quaternary Period. The spectacular alpine scenery of a number of mid-latitude high mountain ranges around the world contains glacier ice today and therefore certainly represents landscapes that have been subject to *average glacial conditions*. These include the North American Rockies, the European Alps, the South American Andes, the Himalaya and Tien Shan mountains, and the New Zealand Alps.

Larger mountain icefields or ice caps exist today in the low to sub-Arctic locations of Iceland, Alaska, and Scandinavia, as well as the higher Arctic and Antarctic islands such as the Svalbard Archipelago, Novaya Zemlya, Severnaya Zemlya, Baffin Island, the Canadian Queen Elizabeth Islands, South Georgia, and the South Sandwich Islands. Additionally, many other mountain ranges around the world display clear signatures of former glacierization but host no ice today.

In all of these locations the extents of multiple phases of more extensive ice coverage beyond the present-day glacier limits have been widely mapped using glacial landforms since multiple glaciations were first formally recognized on the northern foothills of the European Alps by Penck and Brückner in 1909. The compilation of such evidence allows us to reconstruct the changing style of glacierization over time, for example the change to and from niche and cirque glaciers to icefields or ice caps and in some cases to ice sheets. Particularly well preserved are landform records of recent ice oscillations that postdate the LGM and relate to the phase called Neoglaciation.

Glaciers and sea level

Prominent in the media recently has been the mounting awareness of melting glacier ice and its impact on sea level rise. This can be put into perspective by assessing the contribution of our largest present-day ice sheets to sea level rise if they were to melt completely. The Greenland Ice Sheet contains 10 per cent of the Earth's total freshwater and this equates to 6.5 m of sea level rise. More importantly, the Antarctic Ice Sheet is estimated to hold up to 90 per cent of the Earth's total freshwater, which would bring about a sea level rise of up to 60 m if released by melting. However, the West Antarctic Ice Sheet appears to be the most vulnerable to imminent break up, due to its location on a landscape that lies predominantly below sea level; this would bring about a 5 m global sea level rise, an amount still potentially catastrophic for low-lying, heavily populated countries.

Most perceptibly, the concept of lowering ocean levels in response to ice sheet build-up was proposed by American geologist Charles Whittlesey (1808–86), who proposed that sea level should fall by 130 m during global glaciations. We now fully appreciate that the contribution of glacial meltwater to rising global ocean levels is part and parcel of long-term glacial and interglacial cycles, and significant oscillations of global sea level have been measured using tectonically uplifted coral reefs on tropical coasts; as ice sheets grow during a glacial stage, global sea level falls, but conversely, during deglaciation, the melting ice sheets return the meltwater to the oceans to create sea level rise. This process is known as *glacioeustasy* and, as Whittlesey had estimated, amounts to total global sea level changes of 100–120 m.

Although sea level will fall as glaciers and ice sheets form, a further complication in the sea level response to glaciation is introduced in those areas where the ice masses are concentrated. The evidence for this has always been clear to those who inhabit the areas formerly covered by thick ice sheets. This is nicely communicated in the journal of the polar explorer Charles Francis Hall (1821–71) where he recounts a conversation with one of the Inuit of the central Canadian Arctic, who told him: 'Innuits all think this earth covered with water—did you never see little stones like clams and such things as live in the sea up on the mountains?'

Indeed the sea had inundated this region as the direct result of glaciation. Ice is heavy enough when contained within an ice sheet to load the Earth's crust and displace the underlying mantle. This creates areas of crustal depression in a process known as *glacioisostasy*. As a result, those areas that host the largest ice masses can be depressed by several hundred metres below their normal position but will not rebound or recover quickly immediately after deglaciation. This means that they become flooded by the sea even though the global sea level has dropped. As deglaciation progresses, the global sea level will start to rise once more, leading to greater amounts of flooding of the freshly deglaciated and depressed landscapes. Very slow crustal rebound eventually results in sea levels once again dropping. This balance between glacioeustatic and glacioisostatic sea level trends is a complex one that varies in space and time according to a site's proximity to a former ice mass as well as the thickness of the ice mass (see Box 5). The highest total rebound values can be

Box 5 Relative sea level curves of deglaciated landscapes

The idea that marine deposits and shorelines could be raised relative to modern sea level by the rebound of the crust after its loading by glaciation is credited to the Scottish geologist Thomas F. Jamieson (1829–1913). He had identified marine deposits lying above present sea level in the Forth Valley.

The role of glacioisostasy is significant in terms of the record of glaciation, because it results in the emergence of significant areas and thicknesses of glacimarine deposits, deltas, and shorelines (collectively called raised marine deposits) by varying amounts as determined by the location of sites in relation to the thickest ice. For example, around Hudson Bay, the location of the centre of the former Laurentide Ice Sheet, raised marine deposits and prominent flights of shorelines extend up to 250 m above present sea level.

The highest altitude of these raised landforms and sediments in an area is called the marine limit, and the range of altitudes and ages of the marine limit over a region can be used to date the ice recession pattern as well as reconstruct the patterns of glacioisostatic rebound. Similarly, the pattern and age of sea level fall in deglaciated landscapes can be reconstructed using *relative sea level curves* from a range of locations with varying ice thicknesses and deglaciation ages. A relative sea level curve is simply a graph plotting sea level altitude against age, with the oldest and highest raised marine feature (marine limit) being formed immediately after ice recession.

The principles and landform record of glacioisostasy and relative sea level change are depicted in Figure 14. Panel A shows the depression of the crust beneath an ice sheet and the associated formation of a peripheral depression and a forebulge due to the flow of mantle away from the centre of loading. Panel B shows the glacioisostatic signature of a hypothetical island ice sheet at 10,000 years ago and the same, deglaciated island at present, with the pattern of glacioisostatic rebound plotted as isobases based on the altitudes of raised shorelines. Panel C shows an equidistant shoreline diagram, which is a graph with its x-axis aligned at right angles to the isobases and plotting tilts of shorelines of different ages. The letter D demarcates the location of the uppermost and oldest delta or the marine limit. Panel D shows a relative sea level curve for Hudson Bay at the centre of the former Laurentide Ice Sheet, where maximum glacioisostatic rebound has taken place since deglaciation.

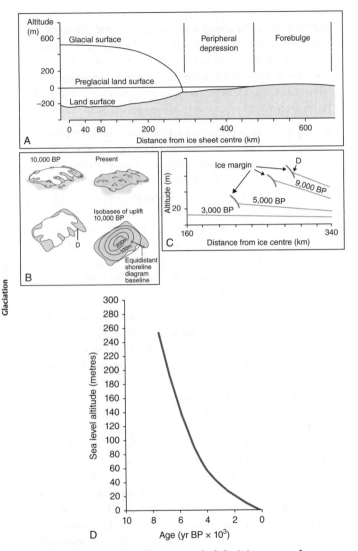

14. The principles and landform record of *glacioisostasy* and sea level change.

contoured to produce *isobase* maps, which provide immediate impressions of the loading centres of former ice sheets.

Historical glacier fluctuations

The most obvious glacial advance episode recorded not just by glacial landforms but also by human documentary evidence is that of the *Little Ice Age*. First acknowledged and named by Dutch-born American geomorphologist François E. Matthes (1874–1948), this was a glacial readvance initiated by cold climate conditions over the period 1500–1900 AD; prior to his studies of the mountain glaciers of the western USA, such ice masses were regarded as remnants of the last glaciation.

Visitors to accessible and tourist-friendly modern glacier margins all over the world will almost certainly be made aware of the evidence for the recent ice-marginal recession that marks the end of the Little Ice Age and the advent of historical global warming. This often will be facilitated by a walk to a glacier snout through inset sequences of moraines, each one adorned with an age marker so that the walker can stop and ponder the fact that the ice was here in the year that they were born.

The Little Ice Age of 1500–1900 AD is but one of numerous cold phases that characterize global climate change since the LGM, each of which has been responsible for positive glacier mass balance and advance and hence what we term Neoglaciation. Prior to 1500 AD was the Medieval Warm Period and prior to that the cold climate of the Dark Ages. Each cold climate snap is now termed a Little Ice Age Type Period, the glaciological impacts of which are recognizable in closely spaced inset sequences of moraines marking the former margins of glaciers in response to each phase of positive mass balance.

On the shortest and most recent of timescales we have records of individual glacier responses to climate change in the form of *mass balance records*. To make direct mass balance measurements on glaciers, especially in the mid-20th century, used to involve painstaking hard work in accessing glacier surfaces on an annual basis, and hence records that extend over ten years or more are available on a mere eighty-six glaciers worldwide. The longest records are for Storglaciären, Sweden (since 1946, Box 6), Storbreen, Norway (since 1949), South Cascade and Blue Glaciers, USA (since 1956), and Peyto, Place, and Sentinel Glaciers, Canada (since 1965).

As the number, quality and frequency of aerial photographs and satellite images have improved over the last twenty years or more, mass balance is being more increasingly undertaken via remote

Box 6 Hans Ahlmann and the Storglaciären mass balance record

Hans W. Ahlmann (1889–1974) was the Chair of Geography at Stockholm University from 1929, taking a leave of absence to become Sweden's ambassador to Norway from 1950 to 1956. He is widely credited with initiating the modern period of systematic glaciological monitoring, also known as the 'Ahlmann period' of 1931–52. Based upon his widespread polar research, Ahlmann identified a strong pattern of glacier recession around the margins of the north Atlantic Ocean, from which came his seminal work in 1948 entitled *Glaciological Research on the North Atlantic Coasts*.

In order to satisfy what he regarded as fundamental needs for long-term monitoring of a specific glacier and bettering our knowledge on climate–glacier mass balance relationships, he identified Storglaciären, in the Kebnekaise massif, Sweden for ongoing measurements. Together with his student Valter Schytt, he thereby initiated the precursor to the modern day Tarfala Research Station where global glaciology's most renowned glacier, Storglaciären, continues to be monitored (Figure 15).

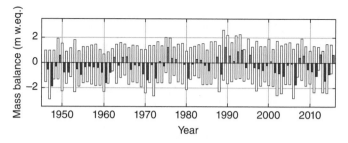

15. **Mass balance record for Storglaciären, 1946–2015. White bars above zero line are the winter balance and below zero line the summer balance. The net balance is denoted by grey bars and is either positive (above zero line) or negative (below zero line).**

assessments of glacier volume over time, no longer requiring the maintenance of networks of ablation stakes on glacier surfaces. The World Glacier Monitoring Service now collects, collates, and publishes all forms of glacier mass balance so that we have a readily accessible, almost real time, inventory of the state of health of the Earth's glacier cover.

Chapter 4
The glacial dirt machine

Glaciers and ice sheets create landscape change in three main
ways: erosion, deposition, and deformation. Erosion includes
direct glacial erosion as well as *glacifluvial* or meltwater-related
processes. Indeed the role of meltwater in glacial landform and
landscape production is often overlooked, despite being capable
of effecting the fastest and most dramatic erosional features.
Deposition similarly involves direct emplacement of material by
glacial ice and meltwater sedimentation, the latter including lake
and marine environments (*subaqueous* processes) and often
dominating large areas of glaciated terrain. Deformation takes
place not only at the ice–bed interface (subglacially) but also
wherever glacier ice moves against soft or weak materials,
compressing and bulldozing them into tectonic landforms
akin to fold mountains in miniature.

The combination of these processes has given glaciers the
reputation of being phenomenally efficient changers of
landscapes. But in reality they modify the Earth's surface only
gradually, eroding and moving material in stages of various
duration via a complex spatial and temporal mosaic of processes,
more akin to a dirt machine or a leaky, restless, and constantly
shuffling conveyor belt.

Glacial erosion

The direct impact of glaciers on their beds was noted well before the glacial theory had taken hold in Europe, with Jean de Charpentier in 1835 writing 'We know that the glaciers rub, wear and polish the rocks with which they are in contact...they follow all the sinuousities, and press and mould themselves into all the hollows and excavations they can reach, polishing even overhanging surfaces'. This observation was quickly followed by those of J. D. Forbes, who in 1843 recovered a sample of basal ice from the Brenva Glacier and described its lower surface as 'set all over with sharp angular fragments...which were so firmly fixed in the ice as to demonstrate the impossibility of such a surface being forcibly urged forward without sawing and tearing any comparatively soft body which might be below it'.

By 1887 the erosive role of ice was summarized by Archibald Geikie (1835–1924) in his textbook *The Scenery of Scotland* in the line 'Pressing onward the sand and stones that lie between it and the rocks over which it moves, it is a powerful grinding machine, that wears down, smooths, polishes and grooves even the hardest rock'. The impact of this erosion was then scaled up to the ice sheet level in 1888 by T. C. Chamberlin in his seminal memoir *The Rock Scorings of the Great Ice Invasions*. These early observations were predominantly related to scratches or *striae* and *grooves* and hence the specific process of abrasion. But other features were evident to Chamberlin, including crescentic-shaped fractures and gouges and *chattermarks*, which he likened to the products of vibratory motions created when cutting tools were forced across hard surfaces.

Chamberlin's musings turned out to be extremely prescient, as we now understand that the subglacial erosion of rock surfaces can be subdivided into two distinct processes of *abrasion*, or the grinding of fine-grained material, and *quarrying*, or the failure of larger pieces or rock due to their being effectively plucked from the glacier bed. Abrasion involves the wear of rock surfaces by the processes of *striation* or the scoring/scratching of bedrock, as well as *polishing*, or the reduction of small-scale rock roughness by the removal of small bed protuberances. Striations (or striae) form where the sharp edges or asperities on rock particles (*clasts*) carried in the ice are dragged over bedrock surfaces, whereupon they scratch tiny grooves (Figure 16).

The perfect analogy for this process is the action of sandpaper on wood, a process that eventually rounds the edges or dulls the quartz crystals (clasts) glued to the paper, as demonstrated in a laboratory experiment conducted by the British glaciologist Hal Lister (1921–2010) and colleagues in 1968; indeed, modern theory and laboratory experiments have arrived at the term 'sandpaper friction model'. Sandpaper also ultimately becomes clogged with the wood residue, being analogous to rock flour in a glacial setting and similarly leading to less effective scouring unless there is a

16. Striations on a bedrock outcrop in Iceland.

mechanism for cleaning out the residue; at a glacier base this is the function of meltwater films, which drain into proglacial streams. Striae are also produced on individual clasts due to their being repeatedly brought into contact with one another in what is essentially a grinding mill. Like all grinding mills the material that goes through it will ultimately be crushed to its smallest possible size, rock flour, and it is this material that is responsible for the dirty, grey-coloured water in glacial meltwater streams.

Rock flour in rivers is diagnostic of a glacial meltwater provenance for the water and as a material in suspension it is responsible for the turquoise colouring of lakes in glacierized landscapes. But before it is released to meltwater it is a material that characterizes a mature or well-developed subglacial sediment called till (see Chapter 5).

Till is typically composed of a whole range of particle sizes ranging from boulders down to clay and hence is described as poorly sorted, but the longer a till has been developing at the ice–bed interface the more crushed will be its components and hence the more clay-rich it appears to become. This was quantified in a seminal study by the pioneer of the modern era of till research, the Latvian-born and Canada-based glacial geologist Aleksis Dreimanis (1914–2011) and his colleague Uldis J. Vagners in 1971.

Testing the general observation that till particle size distributions should be predictable, based upon the fact that they are a product of progressive reduction (*comminution*) due to fracture during shearing, Dreimanis and Vagners sampled grain sizes in tills at progressively greater distances down ice flow in ancient till sheets. They found that finer particles increased in importance at distances of 75–500 km and that there is a lower size limit beyond which no further comminution takes place, regardless of the transport distance. This limit they termed *terminal grade* and is specifically the particle size range from coarse to fine silt; hence early notions that tills were 'boulder clays' were misinformed and any till that contained appreciable amounts of clay must have been created (i.e. deformed) from a pre-existing deposit by glacial overriding.

Quarrying involves the fracture and removal of rock fragments from the glacier bed. The process involved in crescentic gouge production intrigued the pioneer of geomorphological process experiment, Grove Karl Gilbert (1843–1918), who in 1906 proposed that gouges formed where large boulders in the basal ice transmit pressure to the bedrock surface, which then ruptures along a cone-shaped fracture. Experiments on the quarrying process were later reported by Stanley E. Harris Jr in 1943 and Paul MacClintock (1891–1970) in 1953, which involved the scoring of glass with instruments such as files, knives, and ball bearings.

This is a product of stress concentrations created below clasts as they are dragged across the bed, but it is the occurrence of existing cracks in the rock that are most influential in the process. These cracks effectively isolate blocks from the parent rock mass and are essential to the removal of larger fragments than those typical of the abrasion process. Each block that is dislodged then becomes a fresh tool for use in the operation of the subglacial grinding mill. Although abrasion and quarrying are traditionally separated as glacial erosion processes, recent experiments and microscale investigations of striae, such as those by Neal R. Iverson of Iowa State University in 1990–1991, have revealed that each individual striation comprises an elongate chain of chattermarks.

At much larger scales of quarrying, the plucking of large bedrock blocks and ultimately the creation of larger glacial erosional forms, ranging from roches moutonnées (tens of metres, Box 7) to cirques (hundreds of metres up to kilometres, Box 13) to troughs and fjords (tens to hundreds of kilometres), requires significantly greater rock removal. This was expressed by François Matthes in 1930 in his reflections on the origins of the iconic glacially eroded landscape of Yosemite (Figure 17); his

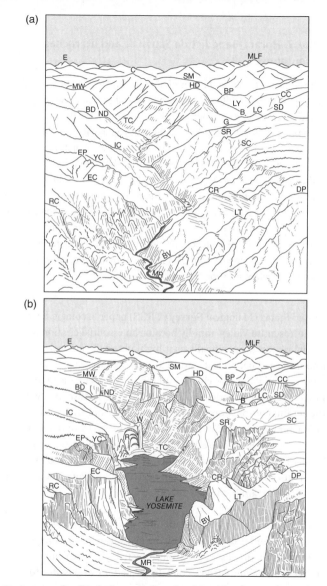

17. An example of the before and after sketches of the glacial modification of Yosemite drawn by François E. Matthes (1874–1948). The labels denote places of local interest, the best known of which is Half Dome (HD).

Box 7 Horace-Bénédict de Saussure and his roches moutonnées

Both the processes and forms of abrasion and quarrying are well illustrated by a landform common to glacial erosional landscapes and called a *roche moutonnée*. This is an asymmetric bedrock bump or small hill comprising an abraded up-ice or stoss side and a quarried down-ice or lee side. Such features were first named by de Saussure, because he thought that they resembled the wavy wigs of the 18th century, called moutonnées after the mutton fat that was used to hold them in place. However anachronistic it now may seem, it is a term firmly entrenched in glacier science and one that is familiar to all students of the subject, who are often taken to their nearest roche moutonnée in order to be shown the impacts of direct glacial erosion all in one convenient location (Figure 18).

United States Geological Survey (USGS) paper 'Geologic History of the Yosemite Valley' rapidly became an essential citation in glacial research, specifically as it highlighted the critical importance of bedrock structure in the selectivity of glacial plucking and landscape change. Into the modern era, the quarrying process has been modelled and experimentally observed, with the seminal outputs on the subject coming from University of Washington geologist Bernard Hallet in 1996 (also renowned for his treatment of the abrasion process and hence the term 'Hallet friction model') and Neal Iverson in 1991, followed by observations from the Norwegian Svartisen Subglacial Laboratory by Iverson's team in 2006.

The Svartisen laboratory lies 200 m beneath the surface of Engabreen and was produced by the excavation of a drainage tunnel for a hydroelectricity project. By fixing an artificial rock step with a shallow surface crack to the bed it was possible to observe the crack growing at times of falling water pressure

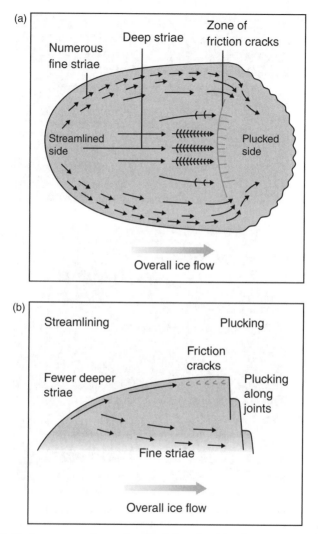

18. Roches moutonnées: a. simplified plan view (a) and cross profile (b) of a roche moutonnée.

(c)

(d)

18. Field appearance of a typical roche moutonnée (c) and the monument in Chamonix (d) of de Saussure and mountaineer Jacques Balmat.

(i.e. when the ice was not being lifted off its bed). Subglacial experiments on glacier erosion such as this are relatively rare. Other notable examples have been those by Geoffrey Boulton in the 1970s at the Salieckna Glacier in Sweden, Breiðamerkurjökull in Iceland, and with Robert Vivian at the Glacier d'Argentière in France.

Meltwater derived from the melting of glacier ice is fundamental to the operation of a glacier as a system, and its role in glacier flow, especially sliding at the ice–bed interface, is also critical to the erosion process. Additionally, some of the most impressive and widespread impacts of glacial erosion are products of concentrated meltwater incision. The significant role of meltwater erosion was demonstrated most remarkably by J. Harlen Bretz (1882–1981) in a case study that geomorphologists generally found to be too fantastic to believe (Figure 19). In trying to understand the evolution of the landscape known as the Channelled Scablands in the US

19. Dry Falls giant cataract alcoves complex in the Channelled Scablands. The dry waterfalls are located in the middle distance and appear as two alcoves or horseshoe-shaped cliff faces. The former plunge pools are now lakes located at the cliff bases.

Midwest, Bretz proposed that the immense size and extent of the erosional features could be explained only by the catastrophic release of huge volumes of glacial meltwater. The origins of this floodwater were related to a vast proglacial lake dammed up in the topography by the advancing southern margin of the North American Cordilleran Ice Sheet during the last glaciation and named by Joseph T. Pardee (1871–1960) as Glacial Lake Missoula.

We now appreciate that there were numerous floods from Glacial Lake Missoula which were controlled by the interaction between the oscillating ice sheet margins and the repeatedly emptying and refilling lake. Each event involved the catastrophic release of up to 2,184 km^3 of water, the equivalent of Lake Ontario. Such events (types of jökulhlaup) are now routinely called *glacial lake outburst floods* (GLOFs) and have been observed at first hand at the margins of Icelandic glaciers, especially during the 1996 jökulhlaup triggered by geothermal activity beneath Skeiðararjökull.

The erosional features, or Channelled Scablands, created by the drainage of meltwater from Glacial Lake Missoula into the Columbia River and out to the Pacific Ocean are of spectacular size and include anastomosing flood channels and gorges with rock basins, giant cataract alcoves, large residual 'islands' or scabs, mega-bars of gravel, and giant current dunes covering an area of approximately 40,000 km^2 (Figure 19). Water depths in the main channels reached 100–300 m and some channels were enlarged by the headward retreat of giant waterfalls, the most celebrated of which is Dry Falls at more than 5 km wide and 120–130 m high. These dimensions have prompted the classification of the Channelled Scablands as one of the geological wonders of the world.

It was the unusual size of these landforms that brought immediate scepticism of Bretz's ideas from the geomorphological research community when he first published them in 1923. Not until the 1960s did the Missoula floods become firmly established as the prime example of the effects of glacial meltwater erosion and it was not long before similar features were recognized in numerous glaciated landscapes. The Missoula flood case study thereby became a prime example of what the eminent geomorphologist William Morris Davis (1850–1934) called an 'outrageous geological hypothesis', challenging accepted thinking and prompting a switch into a new scientific direction or paradigm.

The Missoula floods are examples of *high-magnitude and low-frequency events*, achieving maximum landscape change during only short periods of time. The extraordinary discharges of $2.7\text{–}21 \times 10^6 \text{ m}^3 \text{ sec}^{-1}$ would have been the equivalent to 2–20 times the mean flow of all the world's rivers into the oceans and so the intensity of erosion was very high. Less spectacular are the daily erosional processes associated with normal meltwater production in glacial systems, or the *low-magnitude and high-frequency events*. Such processes are responsible for the smaller-scale incision of the landscape by meltwater running at the glacier bed to form subglacial channels or at the glacier margin to form lateral channels. Nevertheless, significantly large channels can also be cut in such locations wherever meltwater is concentrated over longer periods of time, especially in subglacial environments.

The amount of sediment moved by low-magnitude/high-frequency events in glacial meltwater streams is very difficult to quantify and requires the construction of some significant instrumentation in the stream channels, which the field researcher has to accept might be destroyed at any time during the measurement period. A benchmark study of this sort was conducted on the proglacial stream of Nigardsbreen in Norway in the summer melt season of 1969 by the rigorous Norwegian glacier scientist Gunnar Østrem. To measure the bedload transport, he constructed a 50 m-long steel fence to catch all particles larger than 20 mm in diameter and then probed the depth of the accumulated material twice a day. The suspended load was measured from water samples. Obviously particles between 1 and 20 mm were not trapped by this method and so Østrem's calculations of glacifluvial transport were underestimates.

More sophisticated equipment such as bedload traps has improved the accuracy of modern attempts to measure stream transport loads. Nevertheless, they have confirmed the staggeringly high rates of sediment movement demonstrated by Østrem, which were 400 tonnes of coarse bedload and 1,200 tonnes of suspended sediment in little under a month at the sampling fence. By calculating how much bedload was accumulating at the delta that was forming in a lake at the downstream end of the river, Østrem came up with an annual figure of 11,200 tonnes.

Clearly glaciers are delivering large volumes of material to their proglacial streams and hence their erosion rates are significant. Over time glaciers will preferentially erode parts of their bed to produce areas of overdeepening or depressions in the long profiles of their valleys. Each depression records the location of maximum erosion during any one phase of glacier occupancy, hence an overdeepening will mark the approximate position of the long-term equilibrium line during a particular phase of glaciation (Figure 2). As the maximum ice mass transfer and flow velocity is attained at the equilibrium line, an area of maximum erosion occurs at this location on the glacier bed. Down-glacier of this erosional zone, ice velocity and erosion drop off and consequently subglacial deposition increases towards the snout margin, giving rise to a zone of deposition. Hence, over time glaciers will overdeepen landscapes by exploiting pre-existing weaknesses such as large geological faults or by thickening and thereby flowing faster over pre-existing valleys.

Often underestimated in this large-scale glacial landscape erosion is the process of *rock slope failure*, whereby large masses of bedrock fall from valley sides simply due to the force of gravity (Figure 20). This is a process that delivers significant volumes of rock debris to valley floors merely because surrounding slopes

20. A rock slope failure on the surface of Morsarjökull, Iceland, deposited in 2007 and indicative of cliff instability due to steepening by glacial erosion.

become too steep or oversteepened by glacial erosion. It has been suggested that the maximum altitudes of mountain ranges are maintained by a process known as the *glacial buzzsaw*, a concept that invokes erosion rates removing rock at the same rate that it is tectonically uplifted. A large part of this summit reduction is the failure and collapse of oversteepened slopes due to *debuttressing* and the opening of joints and fractures, also known as *dilatation* or *sheeting*. This involves the expansion of freshly exposed rock due to the removal of the confining pressures when surface layers are removed, for example by glacial erosion. Fractures develop parallel to the slope surface, thereby critically weakening the rock mass and making it susceptible to gravitational failure, especially immediately after deglaciation.

Glacial deposition and subglacial deformation

Vast areas of the Earth's surface and its ocean beds have been affected by glacial deposition, well beyond the maximum margins of former glaciers. It is convenient to divide depositional processes into those that operate on land (terrestrial or subaerial) and those under water (subaqueous). Generalized terms have been used to designate glacial deposits on maps of the Earth's surface, including drift, boulder clay, till, and raised glacimarine sediments, but such classifications hide a multitude of complex characteristics and can often be somewhat inaccurate. Nevertheless, their coverage and thickness is significant and a glacigenic classification implies a specific set of former depositional processes. But contemporary glacier beds and floating ice/deep water interface zones are very difficult environments to access, so how much do we really know about these processes?

After the discovery of the subglacial deforming layer in the late 1970s (see Chapter 2) it was clear that direct glacial deposition was inextricably linked to the deformation process, but we should not really include deformation under a compilation of depositional processes because in a strict sense it is not a form of primary deposition but instead the secondary modification of existing deposits. Nevertheless, more than a century of research on the genesis of till has resulted in a plethora of often confusingly overlapping and apparently contradictory, process-related terms for till and associated sediments and it has become common procedure to include subglacial deformation in process-form discussions on glacial deposits.

To keep it simple we can identify three interrelated and cooperating subglacial processes, whose relative dominance may switch in space and time. First, *melt-out* is the release of debris from the glacier as it melts so that the material is deposited in cavities at the ice–bed interface, where it inevitably becomes subject to further reworking. Second, *lodgement* takes place where individual clasts are arrested at the bed due to their *frictional drag*, which eventually overcomes the *tractive force or* shear stress imparted by the ice motion. Third is bed deformation, which is involved in the creation and transport of all subglacial till but is nourished by melt-out and/or *cannibalization* (thrusting, folding, kneading, pinching, dislocation, and rafting) of the substrate. At any stage, individual clasts can plough through the deforming layer or the substrate and as a consequence their forward movement is arrested and they become lodged. The deforming layer may also thicken over time, especially near glacier margins where basal shear stresses drop, thereby leading to glacier sub-marginal till thickening, a process also manifest in the production of push moraines (see Chapter 6).

It has been suggested by geologists that the easiest way to think of the subglacial till-forming environment is as a *fault gouge* or *shear zone* in which material is constantly generated by the brecciation, mixing, and homogenization of material derived from the substrate, with extra material being input from ice melt-out. A till is only deposited once all deformation stops, even though components of the till (i.e. large clasts or boulders) may become lodged at any time (Box 8). Additionally, deformation may cease gradually, from the base of the deforming layer upwards as the basal shear stress drops off, thereby depositing the till in vertical increments.

Quite often the cannibalization process is incomplete at the time glaciation shuts down and hence the deformed material may still display the characteristics of the parent deposits, such as former lake or fluvial sediments, in which the stratification can still be observed, albeit heavily deformed. This type of deposit is called *glacitectonite* and forms part of a continuum that includes, from least to most deformed, non-penetrative

Box 8 Clast fabrics in tills and boulder pavements

The discovery of the apparent alignment of clasts (stones) in tills, as well as the tendency for numerous boulders to occur in horizontal lines or *pavements* in till exposures, is credited to a handful of very observant glacial geologists who were studying glacial materials in the second half of the 19th century. Later systematic and pioneering work on these phenomena was that of American geologist Chauncey D. Holmes (1897–1981), whose seminal fifty-page paper on the subject of fabric, entitled simply 'Till Fabric', was published in 1941.

Holmes employed exhaustive statistical analyses on clasts in tills and discovered that the majority of clast long axes were orientated parallel to the former ice flow direction and possessed low dip angles. It is now widely recognized that as tills are being created by deformation and lodgement in the subglacial shear zone their enclosed clasts are gradually aligned so that their long axes parallel the direction of stress applied by the overpassing glacier.

A statistical analysis of these alignments can be measured and is called the *clast macrofabric*. It comprises a measure of the orientations and dips of the long axes of a sample of clasts (usually fifty) from the sediment. In a deformable medium like till, the clasts rotate into parallelism with the direction of shear and hence gradually align themselves so that their long axes become orientated and gently dip up-glacier.

Boulder pavements were recognized first by Orange Nash Stoddart (1812–92) and reported by him in 1859, after which their characteristic glacially modified clasts were repeatedly noted to be adorned with striae and facetted and appearing bullet-shaped or flat-iron shaped due to subglacial abrasion and fracture; the parallelism of the long axis alignments and surface striae are also diagnostic. Although some geologists think of these pavements as possible lag horizons created by fluvial, aeolian, or wave winnowing during a non-glacial hiatus, it is now understood that they probably represent the concentration of the larger particles in tills during subglacial till transport.

A till, boulder pavement, and clast fabric at Whitburn, County Durham, England are illustrated in Figure 21. The lower panel shows a clast macrofabric stereonet in which both the orientation and dip of a sample of clasts from a till are presented as a three-dimensional plot. This example shows an orientation and dip of clasts predominantly towards the north-west with medium to low dip angles (i.e. the points plot towards the outside of the stereonet and dip angles increase from 0 to 90 from the outer ring to the centre).

glacitectonite, penetrative glacitectonite, and subglacial traction till (see Chapter 5). The term glacitectonite was introduced by British geologist Peter H. Banham in 1977 in trying to describe and explain the complex structures in the glacial deposits of East Anglia, eastern England, previously highlighted as a product of glacial deformation by British Geological Survey officers in the late 19th century (see Box 12). It is defined as 'rock or sediment that has been deformed by subglacial shearing but retains some of the structural characteristics of the parent material'.

Although the subglacial shear zone is operating whenever ice is in contact with deformable materials, tills and glacitectonites are not ubiquitous in formerly glaciated areas. This should prompt us to acknowledge fully the role of meltwater in glacial deposition. Meltwater is significant not just because of its abundance in glacial systems but also because it acts to modify or remove the majority of primary glacial deposits like tills and influences sedimentation patterns far beyond former glacier margins. This process-form regime is referred to as the glacifluvial environment and can be subdivided into subglacial and proglacial

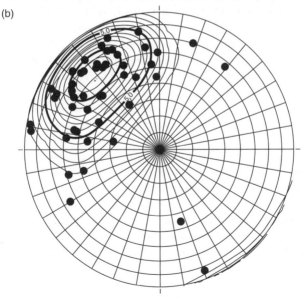

21. A typical boulder pavement in till from Whitburn, County Durham, England (a) and a clast macrofabric stereonet (b).

components where vast amounts of sand and gravel are laid down by running water to form stratified sediment sequences.

In the glacier plumbing section in Chapter 1 the role of meltwater in creating supraglacial (ice-walled) channels, moulins and englacial tunnels, and subglacial water films, cavities or canals, and tunnels was briefly introduced. Such meltwater systems operate in all glaciers and ice sheets to varying extents, dictated by ice dynamics and environmental conditions, but glacifluvial deposits begin accumulating during overall deglaciation, when the meltwater plumbing network gradually enlarges and increases in complexity over time. After studying the appearance of large water-filled holes on Alaskan glaciers in the early 1960s, the American glacial geologist Lee Clayton (1938–2014) equated glacier drainage evolution to karst development in limestone terrains, due to the tendency for the tunnels and cavities to enlarge, increase their connectivity, and eventually collapse; hence the term *glacier karst* has been used to describe the much faster development of drainage networks in downwasting ice.

Important observations on the late stages of the glacier karst production processes were reported in the 1960s by British glacial geomorphologist Robert J. Price (1936–2012), initially on his mapping of *eskers* (winding sand and gravel ridges deposited in ice tunnels and named after the Irish Gaelic term 'eiscir', for ridge) emerging on the foreland of the Casement Glacier in Alaska but then later with a University of Glasgow survey team on the foreland of Breiðamerkurjökull in Iceland. At the latter site it was demonstrated by repeat survey that esker elevations were falling annually and hence they had ice cores that were melting out and so had evolved in englacial rather than subglacial tunnels as traditionally assumed. Because sediment deposition during the evolution of glacier karst systems takes place in a constantly changing environment in which ice can melt-out from around and beneath the accumulating sediment bodies (ice-contact deposition), as demonstrated by Price and his team, the sediments are subject to faulting, folding, flowage, and resedimentation.

These disturbance processes are more effective for supraglacial deposits, which may end up being reworked numerous times due to constant changes in glacier surface slopes. This involves not just the deposits of glacier karst but all debris melting out on the glacier surface and constitutes a process known as *topographic reversal*. Indeed if glacier debris loads are high, the volume of material that eventually melts out on the downwasting glacier surface can form a supraglacial debris blanket that not only retards ablation rates but also results in a ground cover of boulder-rich undulatory mounds (*ice-cored moraine*; see Box 9) and/or chaotic hummocks (*hummocky moraine*; see Chapter 6) once ice has fully melted. It is the production of abundant meltwater during glacier downwasting that normally results in significant reworking of debris into glacifluvial sediments and landforms, termed eskers when deposited in tunnels, and *kames* when deposited in other ice-contact settings (see Chapter 6).

Another significant contribution made by R. J. Price and his University of Glasgow team was the repeat surveying of glacifluvial deposits that had built up to such a depth on the thinning snout of Breiðamerkurjökull that they had formed an outwash fan or *sandur*. The Icelandic term sandur (pl. sandar) is used to refer to spreads of glacifluvial deposits or proglacial outwash laid down by rivers beyond the glacier margin. Often their apexes accumulate over glacier ice and hence tend to collapse after ice melt-out. Price and his team demonstrated that over time the initially flat surfaces of sandur fan apexes collapsed so extensively that they ended up being large areas of chaotic gravel and sand mounds, or *kame and kettle topography*, after being a *pitted sandur* for only a short period of less than ten years.

Box 9 Ice-cored moraine and the Østrem curve

In the winter of 1961, the Swedish glacier scientist Gunnar Østrem conducted an ambitious experiment on the moraines of Isfjallsglaciären in northern Sweden to prove that they contained glacier ice, as he had predicted based upon their surface morphology and resistivity surveys (Figure 22). This entailed transporting pneumatic drilling equipment to the Kebnekaise Mountains and painstakingly excavating downwards through the densely packed rubble.

After many hours of hard labour, proof was handed from the base of the hole to Østrem in the form of a lump of pure glacier ice. He had confirmed that glacier ice could remain buried in the landscape long after deglaciation, but why? We now understand that this is the end product of reduced ice ablation rates, brought about by the protection of the glacier surface by its accumulating supraglacial debris cover.

Very simply, there is a consistent, non-linear relationship between debris thickness and melt rates, as demonstrated by melting experiments undertaken by Østrem, also at Isfjallsglaciären, and now routinely referred to as the Østrem curve. Under very thin debris, ablation rates rise with increasing debris thickness up to a peak value when debris thickness is around 2 cm. Then, under thicker debris, ablation rates decline exponentially.

This pattern is due to two opposing effects. First, in areas of thin or patchy debris cover the lower albedo of rock surfaces compared to bare ice will result in the absorption of more shortwave radiation and an increase in energy for melting. Second, in areas of thick debris cover, the debris acts as a thermal barrier between ice and the atmosphere, reducing the energy flux to the ice surface. Over time, the accumulating debris cover

(*continued*)

Box 9 Continued

will change the strong melting regime to one of ice protection or non-melting. In colder environments the buried ice can lie around for centuries or even thousands of years, effectively turning it into *permafrost* (permanently frozen ground); remnants of the last ice sheets lie buried in the vast permafrost regions of Siberia and northern Canada and are only now being uncovered due to increased permafrost melt in response to recent climate warming.

With the exception of operating often in ice-contact settings, and therefore commonly displaying faulting and folding for example, glacifluvial processes are essentially the same as those operating in river systems (fluvial processes). A significant difference can be identified, however, when comparing the *hydrographs* of glacial and non-glacial fluvial systems and this is manifest in the specific nature of glacifluvial deposits. A hydrograph is a graphic representation of stream discharge over time and for normal rivers it depicts the input of precipitation into a catchment and its influence on stream flow. So individual storms can be identified as rising and falling stages over and above the background discharge and are normally depicted as flood hydrographs. In contrast, the hydrographs of glacial melt streams (Figure 23), because they are fed by glacier ablation cycles, will blend rainfall events with the more dominant signal of glacier melt patterns.

22. a: Gunnar Østrem finally receives proof of his theory that some moraines do indeed have ice cores, as a lump of glacier ice is passed to him from the pit excavated in one of the Isfjallsglaciären moraines in 1961.

b: the Østrem curve (the original curve depicted here for Isfjallsglaciären in 1956) and subsequent variants based upon ablation measurements in a variety of settings. On the Isfjallsglaciären curve, line (a) shows the debris thickness under which maximum melt occurs, and line (b) indicates the thickness at which melt equals that for bare ice.

(a)

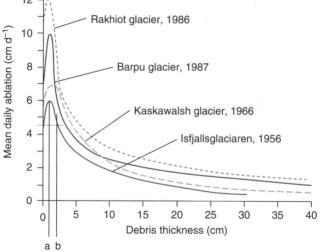

(b)

81

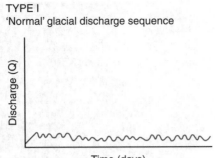

TYPE I
'Normal' glacial discharge sequence

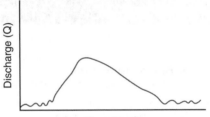

TYPE II
Sudden drainage from
ice-dammed lake

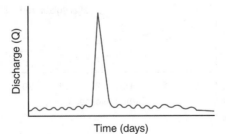

TYPE III
Jökulhlaup triggered by
volcanic eruption

23. **Hydrographs indicative of different types of proglacial rivers.**

Much different and specific only to glacial environments are the hydrographs of jökulhlaups, of which there are those that are related to GLOFs and geothermal (volcanic) activity. Such hydrographs contain flood peaks which are short-lived but orders of magnitude larger than the normal melt-driven cycles. The melt-driven run-off of non-jökulhlaup rivers will tend to be characterized by large amounts of suspended sediment and bedload. They are also characterized by constantly migrating braided channels due to a combination of relatively steep gradients, abundant bedload, cohesionless (erodible) bank and bed material, and fluctuating discharges.

In 1962, Icelandic farmer and self-taught glaciologist Flosi Björnsson (1906–93) published a paper in the Icelandic Earth Sciences journal *Jökull* reporting an event that would have gone unrecorded if he had not been a keen observer of the ongoing glacier processes in his neighbourhood. The event was the catastrophic drainage of an ice-dammed lake at the west margin of Breiðamerkurjökull, leading to the deposition of a spread of glacifluvial gravel and sand over the thin glacier snout. As part of their survey and mapping of the foreland, R. J. Price and the University of Glasgow team were able to do repeat surveys of this fan surface over five years of its early stages of development and thereby demonstrate what Price called in his subsequent paper in 1971 'The development and destruction of a sandur'.

Not only did this provide definitive evidence of the speed of landform evolution in such settings, but it also demonstrated the process-form regime of features common to many glacifluvial river beds, especially near glacier snouts, which are ice melt-out landforms such as pits, craters, or hollows on what should otherwise be flat stream bed surfaces. This gives rise to pitted sandar, and in situations where the pits are very large such features are indicative of glacifluvial deposition over the top of a downwasting glacier snout. Smaller pits distributed over an otherwise predominantly level sandur surface are on the other hand indicative of a different process-form regime, one in which the glacifluvial sediments were deposited along with large ice blocks and hence relate to jökulhlaups when parts of the glacier snout were fractured and carried away in the flood waters.

The huge suspended loads and bedloads of glacial meltwater are not always delivered to the subaerial destinations of sandar or ice-contact landforms such as eskers and kames. Often the meltwater makes its way directly (ice-contact) or indirectly (glacier-fed) to deep water bodies such as lakes or the sea. The processes operating in these depositional environments are referred to as subaqueous and vary hugely according to the proximity of glacier ice. These are identified in schematic diagrams in Figure 24. The upper diagram shows a simplified representation of the sedimentation in ice-contact proglacial lakes or the sea. An englacial tunnel feeds sediment-laden meltwater to the deepwater environment to produce esker-fed deltas or subaqueous fans, which form the proximal depositional zone (A). This passes into the finer deposits of the intermediate (B) and distal (C) zones where some coarser debris may be introduced by iceberg dropping, dumping, or grounding. The lower diagram shows ice-contact and glacier-fed Gilbert-type deltas based on a reconstruction from the Scottish Highlands and shows the typical landforms and sediments of such a setting, including subglacial debris overlying glacitectonically deformed delta deposits, a braided sandur surface, topset beds, foreset beds, and bottomset beds.

A proglacial or ice-dammed lake or the sea can be viewed simply as a basin into which glacier ice feeds sediment either directly via subglacial, englacial, and supraglacial channels or indirectly via proglacial streams, or even a combination of these routes, especially over time. Sediments will be deposited in a spatial pattern dictated by their grain size so that a basin will display *distal fining*, whereby the coarsest grain sizes (gravels and sands) will be

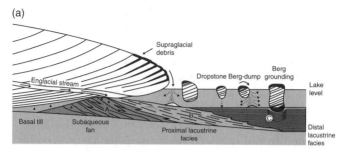

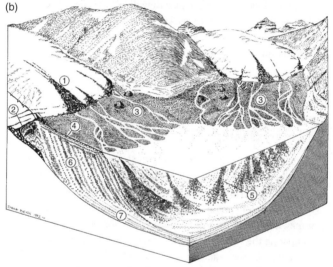

24. **Subaqueous glacial sedimentation styles. a. shows a simplified representation the sedimentation in ice-contact proglacial lakes or the sea. b. shows ice-contact and glacier-fed Gilbert-type deltas based on a reconstruction from the Scottish Highlands, showing (1) supraglacial debris; (2) subglacial debris overlying glacitectonically deformed delta deposits; (3) braided sandur surface; (4) topset beds; (5) delta front; (6) foreset beds; and (7) bottomset beds.**

dropped directly at the entry point, such as at the subglacial tunnel exit in an ice-contact setting or at the mouth of a glacier-fed river.

In contrast, the finer grained sediments (silts and clays) become buoyant and ascend through the water column and move away from the ice margin until they fall through the water column to the lake or ocean floor by a process known as suspension settling. This means that the more distal zones of the lake or ocean are characterized by rhythmically laminated silts and clays, generally referred to as *rhythmites* because they display cyclic repetitions in grain size changes or rhythms dictated by a variety of controlling factors such as glacier discharge changes, seasonal melt signals, and tidal influences (Box 10).

Box 10 Gerard and Ebba Hult de Geer and the discovery of varves

Swedish geologist Gerard de Geer (1858–1943) is widely credited with the discovery that the rhythmically patterned sediment layers laid down in ice-contact and glacier-fed lakes can effectively record annual events. He summarized this in his seminal work *Geochronologia Sueccia*, published in 1940, by comparing the sediments with tree rings: 'From the obvious similarity with the regular, annual rings of the trees I got at once the impression that both ought to be annual deposits'.

De Geer identified that rhythmites deposited in the proglacial lakes at the retreating margins of the last Scandinavian Ice Sheet displayed distinct silt/clay couplets which recorded the marked differences in sediment supply between summer and winter. The coarser silts were deposited in the summer, when higher ice melt and meltwater discharges delivered more sediment to the lakes, and the clays represented deposition from winter suspension settling.

In 1910 de Geer proposed that these couplets be called *varves* and that they could be used to create a geochronology for deglaciated landscapes with a very fine annual resolution. Indeed, his varve chronologies have long served as calibration checks for radiocarbon dating. De Geer and his colleagues collected varve records from many short and overlapping depositional sequences from around south-east Sweden from lakes of various age and thereby created a longer annual chronology of glacial retreat. This varve chronology was later to be called the 'Swedish Time Scale'.

De Geer's identification of annually deposited subaqueous moraines in the Swedish proglacial lakes has also given rise to the adoption of the term *de Geer moraine* for such features. Often overlooked was the contribution of de Geer's wife Ebba Hult (1882–1969), who was effectively his research colleague, travelled with him around the world to study varves and establish chronologies, and with him ran the Swedish Geochronological Institute. After Gerard de Geer's death, Ebba continued to publish his and her own work.

All subaqueous deposits may be greatly disturbed by icebergs as they are subject to drifting across open water and frequent episodes of grounding, scouring, and overturning. Icebergs also drop significant quantities of debris through the water column, the largest clasts falling quickly rather than entering suspension and hence impacting the lake floor deposits and disturbing them to form *dropstones* and *ice rafted debris (IRD)*. This style of subaqueous sedimentation was the one favoured for all glacial deposits by the geological diluvialists before they accepted in the late 19th century that glaciations had taken place on various occasions on the Earth's surface; from such roots came the enduring term 'drift'.

In all of the glacial depositional settings, one set of processes, gravity-induced sediment mass movements or *sediment-gravity flows*, will always be active and often can be dominant. Indeed many sediment-gravity flow deposits can be easily mistaken for primary glacial deposits like tills.

Here we need to introduce an important sedimentological term, whose employment is critical to ensuring the objective and systematic study of sediments in glaciated terrains: this term is *diamicton* or *diamict*, meaning a poorly sorted sediment with a wide range of grain sizes. It is an important descriptive and non-genetic term, because tills and many sediment-gravity flow deposits are all diamictons, often with very similar appearances but produced in very different ways.

The gravitational flowage of sediment–water mixtures creates diamictons by mixing heterogeneous or highly variable sediment sources to create a continuum from stratified (weakly mixed) to massive (strongly mixed) diamictons, often compared to the homogenization process achieved by using a cement mixer. Once common in glacial research was the term *flow till*, which was used to refer to subaerial sediment flows deposited in direct association with glacier ice or from freshly deposited till; similar terms were used for subaqueous glacial deposits such as 'subaquatic flow till' and 'submarine flow till'. A significant flaw in this practice was that once subject to gravity-induced flowage these deposits were no longer actually till per se (Box 11).

Glacitectonics

The deformation of material by glacier-induced stress (glacitectonics) is a subject that forms a sub-discipline in its own right (Box 12). The process of subglacial deformation is critical to understanding shearing processes at the ice–bed interface, but the largest scales of glacitectonic deformation are observed beyond the margins of glaciers and ice sheets where they are referred to as proglacial glacitectonics.

Box 11 Flow till: from birth to retirement

The work of early glacial geologists in the late 19th and early 20th centuries quickly established that glaciers appeared to produce both basal 'lodgement' and supraglacial 'ablation' tills, the latter being involved in flowage and resedimentation after the subaerial release of debris from glacier ice.

The American glacial geologist Joe Hartshorn (1923–2008) proposed the term *flow till* to classify this type of sediment, thereby creating a genetic label for glacial diamictons not created by lodgement or melt-out. The next benchmark was the work by Dan Lawson on the Matanuska Glacier in Alaska in the late 1970s and his proposal that we should subdivide glacigenic deposits into 'primary' and 'secondary' categories; primary deposits are those laid down uniquely by glacial agencies and secondary deposits are those which have undergone reworking by non-glacial processes.

Since that time it has been increasingly recognized that subglacial tills fit the definition of primary, but that all other forms of glacigenic diamicton, because they are remobilized by a combination of gravitational mass flow and fluvial processes, are secondary. Although the term *flow till* still appears in some publications, modern research on glacial sediments has essentially retired Hartshorn's term, preferring instead the more precise but admittedly more convolute terms 'supraglacial mass flow diamicton' or 'glacigenic mass (or debris) flow diamicton' in recognition of their origin as forms of 'sediment flow diamicton'.

Although it is probably evident what the term flow till was trying to communicate (and it was certainly concise), it was nevertheless fundamentally flawed simply because it implied the characteristics of both primary and secondary deposits.

The glacial dirt machine

Box 12 1927, George Slater, and the birth of glacitectonics

In 1927 no less than five papers appeared in various scientific journals on the subject of large-scale glacitectonic deformation, all written by George Slater (1874–1956) of Imperial College London (Figure 25).

In the previous year Slater had introduced the term 'glacial tectonics' in championing a glacial explanation for soft bedrock structures that had been previously traditionally regarded as the products of orogenesis or mountain building. Although Slater has become known as the grandfather of glacitectonics, largely due to his meticulous recording and prolific writing on the subject (eleven papers on moraines and drumlins between 1926 and 1931), he was not the first to propose that glacier ice could initiate large-scale tectonic disturbance.

For example, the glacitectonic disruption of chalk in southern England had been proposed by geological survey officer Clement Reid in 1882 and the British Museum's Henry B. Woodward in 1903, and in Denmark in 1882 by Frederik Johnstrup (1818–94). Also, glacially disrupted bedrock in Alberta, Canada had been reported by Oliver B. Hopkins in 1923. Moreover, the concept of tectonic compression of pre-existing strata had been demonstrated by W. J. Mead in 1925. Modern analogues were also reported by early scientific expeditions to Svalbard in the 1920s and 1930s, exemplified by the eminent German geomorphologist Karl Gripp's work on glacitectonic end moraines or 'stauchmoränen'.

Glacitectonics had fully arrived on the glaciation stage by the 1930s, exemplified by the work of A. Jessen (an early sceptic of the process) and Helge Gry on the spectacular chalk disturbances in Denmark, and J. B. Woodworth and E. Wigglesworth on the well-known complex structures at Martha's Vineyard, Massachusetts.

25. Early field sketch of the glacitectonically deformed materials at the Mud Buttes in Alberta by Slater (1927).

The operation of proglacial glacitectonics was observed during some of the earliest Arctic research expeditions. Of note in this respect was a visit to the Svalbard glacier Sefstrombreen by British glacial geologist George W. Lamplugh (1859–1926) in 1910, while on one of the expeditions led by Gerard de Geer, which Lamplugh then used to provide a modern analogue for ancient moraines on the English east coast. Similarly, in his widely cited 1938 paper on end moraines, the German glacial geomorphologist Karl Gripp (1891–1985) proposed the term 'stauchmoränen' or pushed (thrust) moraines based upon his visits to the thrust and folded blocks that formed the component parts of proglacially thrust moraines on Svalbard during his expeditions in 1925 and 1927.

It was clear to these pioneers of thrust moraine research that the deformation and displacement of material took place in front of the glaciers and was propagated some distance into their forelands. It was not until 1971 that an intensive study of the internal structures and mechanics of a modern push/thrust moraine was reported from the Canadian Arctic by Max Kalin, delivered to the glacial research community via an obscure route as a research report of McGill University.

Despite its rarity, Kalin's report clearly illustrates how thrust moraines develop in a *proglacial stress field*, where the shear stresses imparted by glacier flow are transferred into the foreland. As a result of such proglacial stress fields, impressive compressional structures up to and sometimes exceeding 100 m high are constructed by the folding, thrusting, and stacking of sediments or even soft bedrock in a set of processes akin to fold mountain construction in miniature. This can be demonstrated using laboratory experiments that employ a squeeze box to subject stratified sand to 40 per cent shortening, as illustrated by Figure 26.

Indeed, the production of glacitectonic features has been replicated in laboratory experiments in which the distal stress

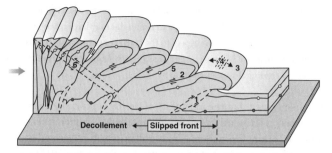

26. An example of piggyback-style thrusting typical of proglacial deformation by an advancing glacier, showing: (1) initiation of a thrust; (2) a thrust fault; (3) a slump zone; (4) extension fractures; (5) a back kink fold; (6) a backthrust zone.

field deposits undergo the full range of compressional and shortening structures seen in fold mountain belts such as overfolds, overthrust faults, piggyback structures, and imbricate thrust slices. The stresses imposed on foreland deposits by an advancing glacier are all that is necessary to create such impressive displacement, although the fact that this is not universal for all glacier margins, modern and ancient, indicates that special conditions must be satisfied for glacitectonic displacement to occur. These include sediment cohesion and confined porewater pressures in the foreland as well as glacier power or driving stress, with surging glaciers appearing to be particularly effective in creating large glacitectonic structures and moraines.

Paraglacial processes

In addition to rock slope failure processes discussed in the 'Glacial erosion' section earlier in this chapter, a wide range of other geomorphological processes are activated and indeed are at their most dynamic immediately after deglaciation. This is especially the case for fluvial environments and drift-covered hillslopes, as identified first by Canadian geographers June Ryder and Michael Church in the early 1970s. They reported:

Glaciation produces a large amount of detrital material that is left in the form of glacial drift. While this material may have reached a position of stability with respect to the glacial environment (that is deposition at the ice margin), it usually is not stable with respect to the succeeding fluvial environment.

They concluded that the instability of freshly deposited glacigenic materials in a recently deglaciated landscape results in their rapid reworking from increasingly gullied, steep slopes and the production of very high fluvial sediment yields immediately after ice recession. This set of processes was termed by them as *paraglacial* and the timescale over which they operate defined as the *paraglacial period*. Paraglacial simply means 'conditioned by glaciation' and occurs at the transition from a glacial to a non-glacial environment.

It is important to investigate paraglacial processes even though they are only conditioned by, and not a primary product of, glaciation because they constitute significant landscape adjustments that would not have taken place if it had not been for glaciation of the catchment. They also create extensive and thick sequences of mass flow diamictons that resemble tills, a problem that has led to much controversy among glacial researchers trying to piece together the jigsaw puzzle of ancient glaciations.

The paraglacial period is characterized by high rates of sediment delivery from slopes and into fluvial and aeolian systems and is a period of rapid response triggered by the instability of unconsolidated glacigenic sediments and oversteepened rockslopes once their support of glacier ice is removed. Rates of change and sediment yields are highest immediately following deglaciation and then decline through time as the sediment supply becomes exhausted and slopes develop more stable profiles.

Steep and unstable glacial landforms, especially if they contain ice cores, will be subject to the most radical changes and will feed

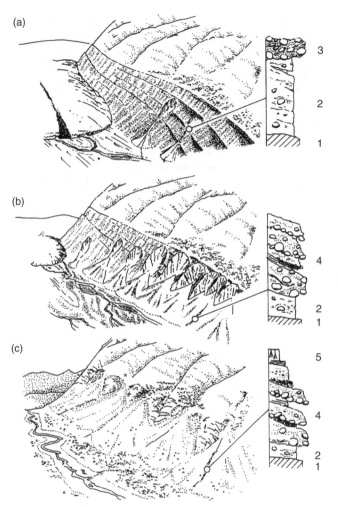

27. Schematic diagrams to show three stages in the development of
landforms and sediments due to paraglacial reworking of steep
moraine covered slopes in a deglaciated valley. The sediment
exposures for each stage include: (1) bedrock; (2) subaerial sediments
relating to an earlier episode of paraglacial sedimentation; (3)
ice-marginal deposits; (4) paraglacially reworked sediment (debris
flows and intercalated slopewash deposits); and (5) soil horizons.

large and extensive sediment-gravity flows until they reach more stable profiles. This is illustrated schematically in Figure 27, which shows three stages in the evolution of a valley side initially blanketed in glacial deposits contained within lateral moraines. At stage (a) the initial slopes have been exposed by glacier recession and are covered by lateral moraines in which gully incision has begun. By stage (b) gully development is more advanced and material removed from the gullies is being deposited in coalescing debris fans downslope. Finally, at stage (c) bedrock is starting to be exposed and the gullies and largely relict debris fans are mostly stabilized and partially vegetated.

Chapter 5
Eroded by ice

There is a very wide range of spatial and temporal scales reflected in the types of glacial erosional landforms, from individual millimetre-wide striae that can form over a few days to fjords tens of kilometres long that require hundreds of thousands of years to develop. Erosional landforms can be discussed in categories defined by three spatial scales, including small or *microscale*, intermediate or *macroscale*, and large or *megascale*, the latter also including whole landscapes which have unmistakable glacial erosional origins.

Although it is convenient to review erosional forms individually, they can rarely be viewed in isolation because they form a superimposed hierarchy in which microscale erosional marks are superimposed on macroscale forms that in turn are superimposed on megascale surfaces to constitute erosional landscapes.

Microscale forms

In the early 19th century, before the glacial theory had taken hold in the geological discipline, a number of early mountaineering geologists remarked on the occurrence of scratches or striation marks on rock surfaces in the European Alps and equated them to the former passage of glacier ice (Figure 16). Indeed, one of the sites traditionally related to the demonstration of the glacial theory by Louis Agassiz in Britain is Agassiz Rock in Edinburgh, on which striae are clearly visible.

A number of erosional forms are essentially these small-scale superficial marks on rock surfaces, recording the passage of single rock particles dragged along by the ice. They include not just striae but also *gouges*, *fractures*, and *chattermarks* as well as an enigmatic set of features termed *P-forms* (Figure 28). Striae are the smallest of all erosional forms and represent the scratch marks of individual clasts, although when viewed closely their floors are composed of numerous fractures or chattermarks, indicating that they have similar origins to the slightly larger forms. Gouges, fractures, and chattermarks comprise either single features or series of features aligned parallel to former ice flow direction, indicating that they were created by repeated fracture events beneath a clast that made intermittent contact with the rock surface as the ice carried it across the site. Single bedrock surfaces can display cross-cutting, multiple generations of these microscale damage trails so that localized changes in ice flow direction can be deciphered.

Macroscale forms

In addition to roches moutonnées (Box 7), macroscale erosional forms include *whalebacks* or *rock drumlins*, *crag and tails*, and *bedrock megagrooves*. Whalebacks are elongate and smoothed, approximately symmetrical bedrock bumps which resemble the backs of whales breaking the ocean surface. The lack of a quarried lee face as found on roches moutonnées is thought to indicate that cavities did not exist at the glacier bed during bedrock abrasion. Crag and tails are elongate, streamlined hills which consist of a resistant bedrock crag at the up-ice end and a tapering tail of less resistant rock extending down-ice. This documents the streaming of ice around the more resistant obstacle and the protection of the 'tail' from erosion. Bedrock megagrooves are relatively rare but important features and record former dominant flow directions by vigorous, likely strongly streaming ice. It is tempting to view them

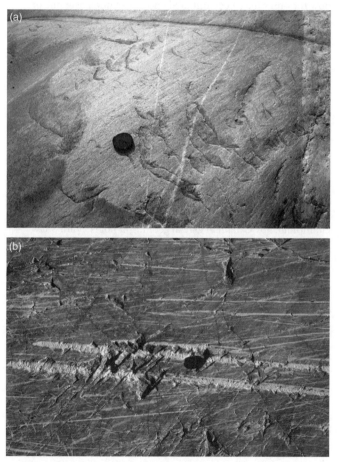

28. **Typical microscale glacial erosional forms: (a) gouges with striae, ice flow from the right; (b) large striae with chattermarks, ice flow from left.**

simply as larger versions of striae but, because they are so large
and have been incised down through solid rock, it is as yet unclear
how they may have been initiated and maintained over one or
more glaciations.

More controversial are P-forms, named by Norwegian geographer
Ragnar Dahl in 1965 based upon his notion that they are produced
by plastically deforming glacier ice. These are smoothed
depressions on bedrock surfaces, often but not always associated
with other microscale forms and sometimes displaying striae
conforming to their curves as if ice had been formerly directed
by their shapes. They exhibit a wide variety of shapes and sizes,
and are usually classified according to their orientation, either
parallel or transverse to ice flow or non-directional. Some P-forms
resemble features created by the fluvial erosion of rocky stream
beds, and hence an entirely subglacial fluvial origin has been
championed by Canadian glacial geomorphologist John Shaw
since the 1990s.

This appears entirely feasible for some features but not necessarily
others. For example, potholes are clearly the product of meltwater
scouring, whereas other features like scallops have been interpreted
as smoothed scars left by the removal of quarried rock fragments.
The controversy remains until we see P-forms evolving in real time,
and so we have to acknowledge that the flow and erosion by four
different media could be responsible for their creation. This includes
debris-rich basal ice, saturated till flowing at the ice–bedrock
interface, subglacial meltwater under high pressure, and ice–water
mixtures. It is highly likely that a hybrid origin is relevant for
many P-forms involving a range of subglacial processes related
to the passage of both deforming till and meltwater.

Megascale forms

Often viewed as the most obvious signature of former glaciation,
the megascale landforms such as cirques, fjords, and U-shaped

troughs represent long periods of glacial erosion of landscapes and indeed demarcate the areal extent of average glacial conditions. Cirques in particular have been instrumental in the understanding and demonstration of glacial erosion, especially since the operation of rotational flow was identified in the 1950s (Box 13).

Box 13 Cirques: glacial erosion in a nutshell

Cirques are often referred to as armchair-shaped hollows cutting back into mountainsides. More specifically the hollow is open downstream and bounded upstream by a steep slope or headwall, which is arcuate in plan form and surrounds a more gently sloping but basin-shaped floor.

This morphology is explained as the product of a combination of subglacial erosion or overdeepening of the floor and subaerial frost action on the backwall, with the most effective glacier for this being one that does not completely fill the cirque. Throughout the first half of the 20th century, geomorphologists argued over the exact origins of cirques and prior to the 1940s the consensus was that they formed by wearing back or headward erosion.

By way of daring and painstaking work in *bergschrunds* and *randklufts* (crevasses at the junction of the upper glacier and the headwall of cirque glaciers) it became apparent that the wearing back was achieved by a combination of freeze–thaw weathering or frost shattering and some glacial plucking. However, this failed to account for the overdeepened basin and it was not until the pattern of rotational flow within a cirque glacier was demonstrated that a mechanism for overdeepening of cirque floors was elucidated.

This breakthrough came about with the Norwegian Jötunheimen expeditions of William Vaughan Lewis (1907–61) and his associates Jean Grove (née Clark, 1927–2001) and J. G. McCall

Eroded by ice

(*continued*)

Box 13 Continued

(1923–54). In 1960 the considerable efforts of Lewis and his team, including students labouring to excavate tunnels into glacier snouts, were published in the book *Norwegian Cirque Glaciers*. This identified the tendency for the glaciers to have wet beds and to rotate, thereby increasing their efficiency in sliding and the basal erosion of cirque floors.

Some calculations of cirque erosion rates (i.e. mature or well-developed cirques) have been attempted more recently and these range from 125,000 years in mid-latitude or temperate locations to 2–14 million years in polar regions, assuming constant occupation by glacier ice. The maturity of cirques relates very generally to the development of their long profiles and their plan forms, which should become more overdeepened and arcuate respectively over time. This concept was first elucidated using headward erosion alone by William H. Hobbs (1864–1953), who was influenced by the *cycle of erosion* advocated by William Morris Davis in which landforms were viewed as representing stages on temporal continuums (i.e. youth to maturity to old age).

A modern take on this is represented by the grading system proposed by British glacial geomorphologist Ian S. Evans, who identifies a range from Grade 1, with all the textbook attributes, down to Grade 5 or a marginal cirque (Figure 29). This approach to reconstructing a sequence of landform evolutionary stages is common in geomorphology where long-term processes cannot be reasonably measured and is called *ergodic reasoning* (substituting space for time).

Prior to this the emphasis was entirely on cirque backwalls wearing backwards only. Once it was understood that cirque glaciers could wear downwards into mountain sides as well as backwards, it became clear how glacial erosion could overdeepen upland landscapes over time. This overdeepening has long been depicted in classic diagrams of mountain glaciation and was initiated by the early work of two of the most famous geomorphologists, William Morris Davis in America and Albrecht Penck in Europe. They identified features such as overdeepened valley floors, hanging valleys, and troughs, all created by the glacial incision of pre-existing fluvial valley networks, a concept beautifully illustrated later by François Matthes in his study of Yosemite (Figure 17).

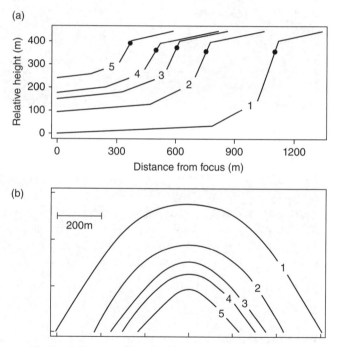

Eroded by ice

29. **The Ian S. Evans grading classification for cirques: a. the grading scheme and its relationship to cirque profiles (a) and plan forms (b).**

(c)

(d)

29. Typical cirques at both ends of the grading scheme with a Type 1 (c) and Type 5 (d), both still containing ice.

Following the lead of Matthes, geomorphologists became more alert to the fact that, at the largest of scales and complexity, whole landscapes could be classified according to their glacial erosion signature. Such landscapes were defined in benchmark papers by British geomorphologists David Linton (1906–71) in 1963 and David E. Sugden in 1968 and 1974 (Figure 30). These landscapes constitute assemblages of landforms created over cumulatively long periods of regional glaciation and include alpine, areal scour, and fjord and trough landscapes.

Alpine landscapes are among the most instantly recognizable landscapes of glacial erosion and constitute networks of troughs and cirques deeply incised into mountain massifs so that summits evolve into ridges or *arêtes* and sharp peaks or *horns*. *Landscapes of areal scour* are extensive tracts of rock knobs, roches moutonnées, and overdeepened rock basins traditionally referred to as *knock and lochain* topography and representing juxtaposed areas of less resistant (lochains or small lakes) and more resistant (knocks or knolls) rocks.

Fjord and trough landscapes are also known as areas of *selective linear erosion*, because they are characterized by deeply incised linear features (troughs or fjords) separated by largely unmodified plateau surfaces and are created by the concentration of erosion by ice streams located in pre-existing valleys. Often still described as U-shaped valleys, troughs and fjords are characterized by very steep bounding walls, but in reality, when their cross profiles were analysed mathematically as early as 1959 by Swedish geographer Harald Svensson, it was discovered that the true shapes are actually parabolas (i.e. the walls are not truly vertical). When viewed on satellite images, the products of selective linear erosion can be seen either as dendritic trough/fjord patterns, where glacial erosion has enlarged pre-existing fluvial valleys, or linear troughs/fjords, where the erosion has been guided by bedrock structures such as regional fault lines.

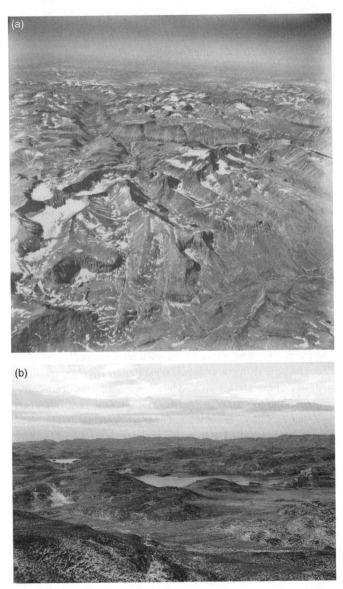

Meltwater erosional landforms

The huge volumes of meltwater created by glacial systems are responsible for the erosion of significant channels at a range of scales from those less than 1 m deep to those that are so large and extensive (e.g. the Channelled Scablands) that they are visible from space (Figure 19). The processes active in the creation of such features are essentially fluvial in nature but, because they are specifically related to the production of meltwater by melting glacier ice, they are called glacifluvial landforms and can be cut either subglacially, ice-marginally, or proglacially.

Subglacial channels cut into bedrock or sediment are called Nye channels (Figure 9), after John Nye, but at the largest of scales they are normally termed *tunnel valleys* or *tunnel channels*. Because the subglacial water in these channels flows under pressure it can flow uphill and across topographic obstacles. Cutting by pressurized water also gives rise to their diagnostic undulatory long profiles with overdeepened basins along their floors, as well as cross sections comprising relatively flat bottoms and steep sides.

Ice-marginal or *lateral meltwater channels* are cut by meltwater that flows over a glacier snout to drain along its margin instead of penetrating to the subglacial drainage network. This is more common in cold-based or polar glaciers and results in the incision of bedrock or sediment on the valley walls that are in contact with the ice. A receding glacier margin can thereby leave a series of inset lateral meltwater channels, allowing glacial geomorphologists to map out its retreat pattern in the landscape (Box 14).

30. **Glacial erosional landscapes: (a) alpine landscape (foreground) and fjord and trough or selective linear erosion landscape (distance) in the Torngat Mountains, Labrador, Canada; (b) areal scour landscape of knock and lochain terrain, north-west Scotland.**

Box 14 Carl M. Mannerfelt and the Scandinavian School

What became known as the 'Scandinavian School' of glacial geomorphology was launched with the publication of a 239-page tome, written in Swedish by Carl M. Mannerfelt (1913–2009) in 1945.

Mannerfelt was introduced to contemporary glaciers while undertaking field research with Hans Ahlmann in Iceland in the 1930s (see Box 6) and hence he was well equipped with the necessary observations on modern glacial processes to make some profoundly perceptive reconstructions of deglaciation and meltwater landform production in his native Scandinavia. From this work came the concepts of spillways cut by decanting ice-dammed lakes, ice-marginal meltwater channels, and subglacial drainage networks. The Scandinavian School was rapidly developed also by the seminal works of Valter Schytt (1919–85), Gunnar Hoppe (1914–2005), and Just Gjessing (1926–2005).

Proglacial meltwater channels are excavated by meltwater once it leaves the glacier system and are at their most impressive when produced by flood waters or GLOFs. As we have seen with the case of the Channelled Scablands and the Missoula Floods, some of the most spectacular glacifluvial erosional landforms are created by catastrophic drainage of glacier-dammed lakes. Where such lakes deepen to the extent that their water level breaches a low point in a watershed they cut channels that are generally classified under the umbrella term 'spillway'. They are conspicuous in deglaciated landscapes because they cross watersheds between drainage basins and hence can be explained only by the filling of one valley with a lake, which in turn requires the valley to have been dammed temporarily.

Valley damming and spillway cutting can result in some significant drainage diversions that then persist after deglaciation.

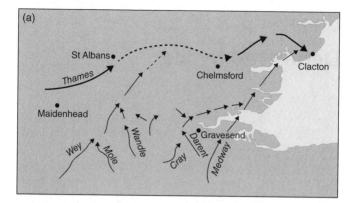

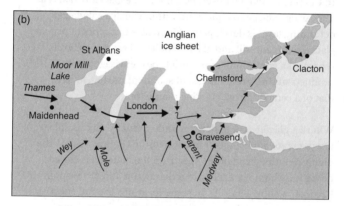

Eroded by ice

31. Maps showing the diversion of the River Thames during the Anglian glaciation at 450,000 years ago.

One of the most renowned drainage diversions is that of the River Thames (Figure 31) which was forced to move from its northern route through the Vale of St Albans to its present, more southerly course. This diversion was initiated after the Vale of St Albans was turned into an ice-dammed lake by the advancing British Ice Sheet during MIS 12 (Anglian Glaciation) around 450,000 years ago.

Chapter 6
Deposited by ice

In Chapters 1 and 3 we established that the glacier bed acts as
a shearing zone or traction zone within which a subglacial
deforming layer is developed. This creates two types of primary
glacial sediment: till and glacitectonite. These two deposits
represent a continuum of materials best envisaged as the
products of the various stages of sediment mixing, akin to
that of a cement mixer, creating diamicton.

Where a glacier acts to deform and gradually mix its substrate
it will create a melange or glacitectonite (Figure 32). At the
highly deformed end of the spectrum the material is close to
being fully homogenized into a diamicton and at this point it
becomes a subglacial till. In some locations the full spectrum
might be preserved as a vertical continuum comprising
undisturbed substrate passing upwards into non-penetrative
glacitectonite with mildly distorted primary structures, then
into penetrative glacitectonite with widespread, penetrative
shear structures and finally into till. This represents the
glacier bed shear zone, the only place where primary glacigenic
material is created and hence not the most widespread of
the glacial deposits when compared to those laid down
by meltwater.

Glacitectonites

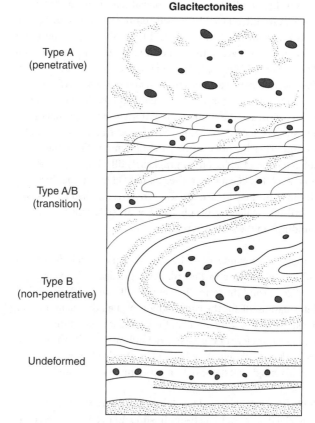

Type A
(penetrative)

Type A/B
(transition)

Type B
(non-penetrative)

Undeformed

Deposited by ice

32. The glacitectonite continuum. The idealized vertical sequence
displays increasing maturity with height so that completely
undeformed parent material is at the base and a penetrative Type A
glacitectonite is at the top.

Till

The term *till*, used by the Scots to refer generally to coarse, stony soil, was first applied to primary glacial sediments in 1863 by geologist Archibald Geikie in a substantial and much underrated publication entitled 'On the Glacial Drift of Scotland', published as the first 190 pages of the first ever volume of the *Transactions of the Geological Society of Glasgow*. The term was nevertheless often overlooked by early glacial researchers, who preferred instead the less appropriate term 'boulder clay'. The variable characteristics of the component materials of this primary glacial deposit were soon to be employed by American geologists in the late 19th to early 20th centuries to differentiate 'lodgement' (subglacial) and 'ablation' (supraglacial) tills. Indeed, as early as 1894, Archibald Geikie's brother, James Geikie, in the third edition of his benchmark text *The Great Ice Age* had identified the tell-tale signatures of subglacial wear (striations and edge rounding) on the stones in till.

Through the second half of the 20th century glacial researchers had arrived at the term 'flow till' for supraglacial deposits (see Box 11) and the terms 'lodgement till', 'melt-out till', and 'deformation till' for the end products of the other main primary glacial processes. Influential in the establishment of this nomenclature for till was Geoffrey Boulton, initially through his work on sedimentation by modern Arctic glaciers and later his subglacial deformation experiments in Iceland.

Boulton's work identified two important concepts: first that englacial debris could melt-out in situ and potentially preserve its englacial structure as melt-out till; and second, that a single glaciation could be recorded in a tripartite till sequence of lodgement/melt-out/flow till. This was demonstrated by Boulton for ancient till sequences using the site at Glanllynnau in North Wales (Figure 33).

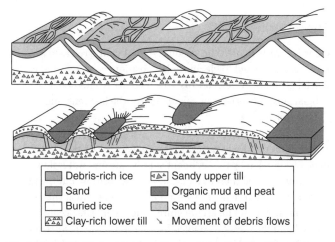

33. A schematic diagram to show how complex glacial sediments in the stratigraphic record (lower panel) can be interpreted using the process-form relationships observed on modern downwasting glaciers (upper panel) and also conveying the principle of tripartite (till-stratified sediments-till) sequences relating to one glacial advance.

His later subglacial experiments then added a further subglacial deformation component to the tripartite scheme. This scheme is still viable today but terminology has changed as our knowledge of depositional processes improves. For example, the term flow till is no longer regarded as appropriate and has been replaced by supraglacial mass flow diamicton. Additionally, all subglacial tills undergo both lodgement and deformation and hence the term 'subglacial traction till' is now used to refer to 'sediment deposited at a glacier sole after release directly from the ice and/or liberated from the substrate and then disaggregated and completely or largely homogenized during transport'. They are therefore dense, compact and often cross-cut by shear planes and display clear clast alignments or fabrics (see Box 8). Melt-out tills, due to the fact that they require very passive and slow melt of their encasing ice and no subsequent flowage (the latter being very rare in glacial settings), have a very low preservation potential.

The 'stratified drift'

Once ice starts to melt the various types of debris released as a result are subjected to glacifluvial deposition, creating what early glacial geologists referred to as 'stratified drift'. This can take place subglacially, subaerially, or subaqueously, the last referring to deep water settings.

In detail there is huge variety in the appearance of glacifluvial deposits dictated by the sediment supply rate and water flow conditions as well their exact location of deposition, but the most common characteristics are abrupt lateral and vertical changes in grain sizes, 'cut and fill architecture', where deposits become incised and the incisions refilled by later flows, and disturbance structures, created by the melting of adjacent or underlying ice, and glacitectonic deformation. Such deposits are created in subglacial settings in Nye channels/tunnel valleys and in ice-walled tunnels, where eskers form, or even in subglacial lakes. In subaerial settings glacifluvial deposits occur in their least disturbed form in outwash plains or sandar and in heavily disturbed assemblages (kame and kettle topography) wherever sedimentation was in supraglacial streams and/or where outwash has overwhelmed a downwasting glacier snout.

The recognition that ancient gravel and sand plains in northern Europe were akin to the modern sandar is credited to the German naturalist and geologist Otto M. Torell (1828–1900), who in 1857 reported on his observations made during a visit to Iceland. However, it was another German geologist, Konrad Keilhack (1858–1944), who in 1883 appears to have been the first person to propose the use of the Icelandic term 'sandur'. At around the same time, North American geologists were describing the stratified drift and identifying ancient sandar associated with the margins of the Laurentide Ice Sheet in the USA. A little later, at the turn of the 20th century, expeditions to

Alaska were allowing American geologists such as Israel C. Russell (1852–1906), Grove K. Gilbert, and Ralph S. Tarr (1864–1912) to observe proglacial outwash being deposited and thereby use this as an analogue for interpretations of the ancient stratified drift further south.

Although the principles of sedimentology were well established by the turn of the 20th century, the complex stratified deposits of former glaciations had been somewhat overlooked, as summarized by J. Kaye Charlesworth (1889–1972) in his monumental catalogue of everything on the Ice Age, *The Quaternary Era*, published in 1957:

> Although the clays, sands and gravels belong to the youngest and most accessible formation, their apparently chaotic state and seeming lack of interest made them the last to be investigated: they were for long a synonym for confusion, and except for their fossil shells and bones seemed unattractive and unimportant.

However, the most observant of the early glacial geologists, for example the Geikie brothers, Archibald and James (Box 3), and Thomas F. Jamieson in Scotland, had described the intricacies of kames and eskers in the 1860s and 1870s. For example, James Geikie in 1877 described the internal sediments of kames as:

> usually stratified, and, in some of the finer grained accumulations, very beautiful examples of false or diagonal bedding frequently occur. But in many cases the coarser heaps of gravel and shingle do not exhibit any trace of stratification, the stones being piled up in dire confusion.

He also described the range of rounding characteristics of the stones, with angular forms being as common as rounded ones, and the clear folding and crumpling of strata related to ice-contact sedimentation.

At the same time in North America the internal stratigraphy of eskers and kames was being described and correctly related to the characteristically non-uniform discharges of glacial meltwater streams. For example, George H. Stone (1841–1917) in studying the eskers of Maine identified what we would now call *palaeocurrents* in the cross-bedded sands and gravels, indicating water flow in the tunnel towards the former glacier snout. He also explained the huge variety in particle sizes as the products of 'moderate currents for most of the time, with now and then a sudden flood'; this simple description of glacial meltwater sedimentation is still entirely valid.

Much of the glacifluvial sediment that is transported to glacier and ice sheet margins ends up being deposited in lakes and oceans. The vast proglacial lakes that developed along the southern edge of the retreating Laurentide Ice Sheet have long been recognized, especially 'Glacial Lake Agassiz', named after the traditionally acknowledged grandfather of the glacial theory. The interaction of subglacial and subaerial (terrestrial) glacifluvial regimes with such subaqueous (deep water lake and marine) settings brings about an abrupt change in grain size characteristics in diagnostic 'depo-centres' called *deltas* or *grounding line fans* where sediments are arranged in *clinoforms* or sequences of steeply dipping beds created by progradation due to the falling discharge of the transporting medium.

A benchmark in the definition of the complex coarse-grained deposits laid down in such settings was the study of Canadian geologists Brian R. Rust and Richard Romanelli in 1975, based on the subaqueous fans laid down by the receding Laurentide Ice Sheet margin into the glacioisostatically raised marine waters in the Ottawa River valley. They pointed out that such deposits were not unlike glacifluvial outwash but, because of their delta-like association with finer grained deep water sediments, they could not be classified as anything other than subaqueous outwash.

The overriding control at this interface is the sudden drop in velocity and hence carrying capacity of the inflowing water so that the resulting subaqueous sediments are spatially arranged according to their proximity to the influx point and/or glacier (i.e. proximal/coarse-grained to distal/fine-grained deposits). This is termed *proximal to distal fining* and is commonly illustrated using the example of a typical *Gilbert-type delta*, named after G. K. Gilbert and common in lakes fed by debris-charged rivers typical of glacial environments (Figure 24). Such deltas, in contrast to Rust and Romanelli's subaqueous outwash fans, are composed of three depositional components including *topsets* or the proximal fluvial sediments of the feeding sandur, *foresets* deposited by gravitational processes on the delta slope, and the more distal and finer *bottomsets* deposited by *underflows* in the deeper and slower flowing water on the delta toe.

At around the same time that G. K. Gilbert was developing his model of delta sedimentation, American glacial geologists were delivering benchmark descriptions of the deposits laid down during the last glaciation in proglacial lakes, for example in Lake Agassiz by Warren Upham (1850–1934), who in 1895 delivered the earliest tome on the lake, entitled *The Glacial Lake Agassiz*, which was to be the catalyst for hundreds of research papers from that point and into the modern era.

Distal subaqueous sedimentation in glacially influenced deepwater settings such as lakes and the oceans is responsible for the most extensive of the glacigenic deposits on Earth. They were identified in glacial lake deposits and then surmised to be related to annual or seasonal sedimentary rhythms by the 19th century work of American geologists Warren Upham and Edward Hitchcock (1793–1864; Glacial Lake Hitchcock in Connecticut was named after him). These deposits are called *glacilacustrine* and *glacimarine* sediments respectively and comprise rhythmically bedded (rhythmites) or laminated fine-grained materials such as fine sands, silts, and clays.

They can form varves where seasonal sedimentation controls the grain size variability, a term that was attached to these deposits by Gerard de Geer (see Box 10), and invariably contain iceberg-related features such as dropstones and ice rafted debris (IRD) and iceberg grounding disturbance structures (Figure 24).

In contrast to glacifluvial and glacilacustrine research, studies of glacimarine deposits were slow to develop, largely because of their far more restricted exposure and the lack of exploration of submarine environments on glaciated shelves. The earliest clear acknowledgement of glacimarine sedimentation, delivered as it was in a scientific climate of diluvial explanations for all glacial deposits, was that of Archibald Geikie in 1863 and his description of fossiliferous stony muds interbedded with marine clays ('marine drift') deposited by icebergs and sea ice. The earliest offshore samples were collected from the sea floor off Antarctica by Emil Philippi (1871–1910) during the German Antarctic Expedition 1901–3. He defined the core samples of sands, cobbles, and pebbles in a finer grained matrix as 'glazialmarine Ablagerungen' or glacial marine sediment.

In the 1950s it was exposures through glacimarine deposits, brought about by their uplift above sea level, which formed the basis of new developments. Working on deposits dating to the last glaciation in British Columbia, Canadian geologist J. E. (Jack) Armstrong (1912–95) described interbedded diamictons with marine clays, silts, and sands containing marine fossils. But the first example of intensive sedimentological investigations of glacimarine sediments came from the much older rock record, with USGS officer Don J. Miller (1919–61) in 1953 reporting pre-Quaternary till-like strata on Middleton Island, Alaska, which he named 'Yakatagite'.

A boom in research on glacimarine deposits came with the increase in deep ocean drilling in the 1960s, but it was the year 1981 that saw the publication of three seminal papers on the nature of 'stratified drift' as glacimarine deposits. First, David J. Drewry and A. P. R. Cooper (Scott Polar Research Institute, Cambridge) produced a process-form model for glacimarine sedimentation around Antarctica. Second, Alan R. Nelson (University of Colorado) delivered a much underrated and underutilized overview of multiple tills and glacimarine deposits, recording the alternation of glacial and marine environmental conditions on Baffin Island. Third, Ross D. Powell (Northern Illinois University) published his model of sedimentation at the margins of retreating tidewater glaciers in Alaska.

At the start of the 21st century the shape of things to come was signalled by Powell's employment of unmanned submersibles to access the grounding lines of glaciers and to verify that subglacial till was extruding from the ice–bed interface and delivering coarse-grained deposits to morainal banks and subaqueous fans.

Erratics

Even before the acceptance of the 'glacial theory' in the latter half of the 19th century, the anomalous occurrence of boulders in locations far from their natural parent outcrops had been noted and considered to be inexplicable in terms of the accepted geomorphological processes at that time. Hence the concept that they must have been drifted in by icebergs during a great flood was generated by the influential geologist Charles Lyell, giving birth to the term 'drift' for the deposits that later became accepted as glacially derived rather than the products of a biblical deluge.

In northern Britain a body of enthusiasts became enthralled by the anomalous boulders that were often put on display in town centres and on village greens (Figure 34) and took on the responsibility of mapping them out and thereby tracing their long-distance transport paths. This North-West of England Boulder Committee reported regularly and eventually more

34. The Shap Granite erratic displayed outside the town hall in Darlington in north-east England. The information board explains to passers-by how this huge boulder was transported by glacier ice from the Lake District and across the North Pennines watershed by easterly flowing ice streams in the former British-Irish Ice Sheet.

permanently through the auspices of the *Glacialists Magazine*, set up in 1893 by British glacialist Percy Kendall (1856–1936), American glacialist Warren Upham, and the president of the new Glacialists' Association, Charles E. De Rance (1847–1906) of the British Geological Survey.

Two of the most prominent of these *glacial erratics* in the British Isles are the Ailsa Craig Granite, transported from the Scottish island of Ailsa Craig southwards through the Irish Sea to south Wales and southern Ireland, and the Shap Granite, carried from the English Lake District and across the Pennines to the North Sea coast. In Canada, the best-known example is the *Foothills Erratics Train* comprising huge quartzite boulders, the most prominent being located on the high plains south of Calgary, but the whole linear assemblage stretching 900 km from their source outcrop in the Rocky Mountains near Jasper southwards into Montana.

In every one of these and numerous other cases, the erratics represent the final resting place of debris carried by glacier ice, often over long distances and across numerous drainage divides. Although it is tempting to draw lines from source to resting place and thereby imply a glacial flow direction, most erratics have likely been reworked numerous times during more than one glaciation. Similarly, till is composed of a range of materials, some derived locally and some from more distant locations.

Chapter 7
Landforms from the restless conveyor belt

As we have seen in preceding chapters, the study of glacial landforms and sediments has a long pedigree and forms a branch of geomorphological endeavour called *glacial geomorphology*. The term 'geomorphology' means simply the study of earth forms and is approached from two general angles.

First, process studies involve empirical field measurement, survey and observation, and laboratory experiment and thereby employ an ever-expanding knowledge base on glaciology, process sedimentology and numerical modelling. Second, form analogy, which could also be called reconstructive geomorphology, involves the assignment of a genetic classification to a landform or landscape informed by known process-form relationships. By combining these approaches glacial geomorphologists attempt to make sense of the vast array of glacial depositional landforms created by what we termed earlier the leaky, restless, and constantly shuffling conveyor belt or 'dirt machine' that is the glacier.

A useful process-form-based framework in which to pigeon hole glacigenic features is that of spatially defined 'sediment–landform associations', which include subglacial footprints, ice-marginal moraines, supraglacial associations, glacifluvial assemblages, and subaqueous assemblages. This framework also helps us to categorize glacial landforms using the process information reviewed in previous chapters.

The footprints of glaciers

The processes operating at the base of a glacier or ice sheet produce a diagnostic subglacial depositional footprint. Such footprints record the former extent of ice sheets and glaciers and comprise all those landforms associated with the operation of the subglacial traction zone (shear zone) as well as the subglacial meltwater system. Prominent among these landforms are the most enigmatic and often controversial of glacial features called *drumlins*, a name derived from the Gaelic term 'druim' or rounded hill.

Drumlins are smooth, oval-shaped or elliptical hillocks with long axes orientated parallel to former ice flow and occur in *swarms*. In 1959 the eminent and widely respected Cambridge geographer Richard J. Chorley (1927–2002) published a paper that proclaimed to have derived the mathematical shape definition of a drumlin to be that of a lemniscate loop or teardrop shape. This was clearly predetermined by the traditional definition of drumlins as resembling 'baskets of eggs' in the landscape and was not actually based upon a significant sample size. But the paper endured for far too long, as in essence it was a red herring, failing as it did to encapsulate the huge variety in drumlin shapes that had been mapped worldwide up to that time.

Drumlins can be classified according to their elongation, measured simply by dividing their length by their width and thereby deriving an *elongation ratio* (ER). This reveals that a continuum of forms exists from almost round or ovoid features, through teardrop shapes (traditionally but also mistakenly regarded as the typical drumlin form), to spindle shapes. At the more elongate (spindle) end of this continuum, the term drumlin is replaced by the term *megaflute* and this in turn by *megascale glacial lineation* (MSGL).

Other prominent landforms in the subglacial footprint are *ribbed terrain*, also termed *Rogen moraine* after their initial discovery in the Lake Rogen area of Sweden; these features lie transverse to ice flow but display arms that bend in the down-ice direction.

The internal sediments and structures of drumlins, megaflutes, MSGL, and ribbed terrain are varied and may comprise till from top to bottom or merely a till carapace over the top of a stratified sediment core. Debates continue over the origins of these landforms but what is clear is that they all document smoothing or streamlining of the glacier bed and provide an immediate indication of former ice flow direction; hence they are known as *subglacial bedforms* and are often referred to as the *subglacial bedform continuum* (Figure 35). The continuum is a concept that was first elucidated by the British glacial geomorphologists David E. Sugden and Brian S. John by drawing upon the vast number of observations of glacial geomorphologists working on subglacial bedforms of the Laurentide and Fennoscandinavian ice sheets.

Ice velocity (m a^{-1})

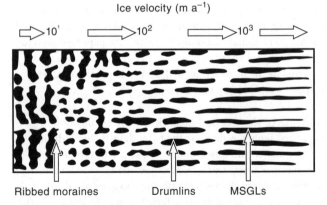

Ribbed moraines Drumlins MSGLs

35. **The subglacial bedform continuum. Here the spatial change is related to ice flow velocity whereby more elongate forms are indicative of faster ice flow, but the changes to more elongate forms can also be related to the length of streamlining period.**

So how do glaciers and ice sheets create subglacial bedforms? This is a question that has vexed glacial researchers for more than a century and from their deliberations on this problem a plethora of theories have emerged, some more fanciful than others. Like any transporting medium such as flowing water or the wind, the answer ultimately lies in understanding the processes that operate in the 'traction zone' or the ice–bed interface. Two sets of experiments, presented earlier, are important in this respect in that they implicate subglacial deformation as the prime contender: first, the subglacial deforming layer experiment conducted by Geoffrey Boulton beneath Breiðamerkurjökull in Iceland, which demonstrated that a deforming till layer operates beneath modern glaciers; and second, recent geophysical studies on the beds of Antarctic ice streams, which have revealed rapidly changing subglacial bedforms (MSGL and drumlins) that appear to be migrating downflow by incremental plastering of deforming till.

These findings beneath modern ice sheets are a vindication of a group of perceptive glacial geomorphologists who worked on drumlins in the late 19th and early 20th centuries, not least Herman Le Roy Fairchild (1850–1943), who in 1907 published a beautifully illustrated and eloquently written report on the drumlins of New York State, USA, proposing their production by plastering-on of till at the ice–bed interface. Also important has been the development of an understanding of the smaller-scale version of flutes on modern glacier forelands, especially as they have been reconciled with the subglacial deformation experiments operating under a glacier snout that is actually creating minor flutings (i.e. Breiðamerkurjökull). Instantly recognizable, they comprise relatively evenly spaced, parallel ridges up to 3 m high and wide, giving the impression of a ploughed field in front of the receding glacier. They are composed of till and often start at lodged 'stoss boulders'. The clast fabrics in the flutes reveal that the deforming till is squeezed from areas of high pressure into relatively low pressure lee side cavities on the downflow sides of lodged boulders, and once in place the till propagates the cavity so that the fluting grows downflow. Flutes without stoss boulders suggest that grooves might be ploughed by the boulders while they are dragged through the till prior to lodging, a process that has also been proposed as the origin for MSGL but yet to be fully tested.

So a process-form regime has slowly emerged to explain the streamlining and/or smoothing of subglacial bedforms, but why the form continuum? The answer to this lies in the recognition that subglacial bedforms belong to suites of different age. This was first identified in drumlin swarms that appeared to contain overprinted or superimposed examples, a concept first elucidated systematically by British geomorphologists Jim Rose and Jocelyn Riley (née Letzer, 1943–2014) in 1977 using the drumlins of the Glasgow area and Vale of Eden.

Once satellite imagery became widely available, the concept of superimposed subglacial bedforms was firmly established, heralded by the seminal study of Canadian geologists Arthur S. Dyke and Thomas F. Morris in 1988 in their identification of the cross-cutting footprints of several former ice stream beds in the central Canadian Arctic. The first ice-sheet-wide studies of such imprints was to follow in the 1990s with the work of British glacial geomorphologist Chris D. Clark, who compiled conceptual models to explain how subglacial bedforms may overprint and partially modify each other, indicating that the ice flow direction changed during bedform production but was often not vigorous or long-lived enough to entirely erase earlier imprints (Figure 36; see Chapter 8). Therein was born the notion that subglacial bedform imprints could be 'smudged', a term that perfectly describes the tendency for glacier ice to more commonly deform rather than erode its bed.

Subglacial bedforms, especially drumlins, have become the most controversial of all glacial landforms, simply because it is impossible to directly access ice sheet beds and observe them evolving. So for those who advocate subglacial deformation

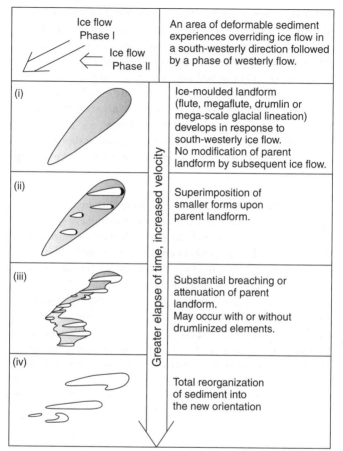

36. Conceptual diagram to show how the shift in an ice sheet dispersal centre can result in cross-cutting subglacial bedforms due to the moulding of subglacial materials by two different ice flow phases. The longer the second flow phase lasts or the faster its velocity, the greater the degree of remoulding of the first phase landforms.

and streamlining of till beds the most exciting breakthrough on drumlin formation came with reports between 2003 and 2009 that British Antarctic Survey scientists had measured drumlins growing and migrating at the base of the Rutford Ice Stream, using seismic and radar techniques repeated over a series of years. When reconciled with the growing database on subglacial deforming till, initiated by the Ice Stream B experiments of the 1980s, it was apparent that landform, sediment, and process could at last be genetically linked at the ice sheet scale and the more than century-old 'drumlin problem' was finally approaching a solution.

Also prevalent in glacier and ice sheet footprints are the depositional records of the former meltwater drainage system, comprising the sinuous sand and gravel-filled ridges called eskers. On ancient ice sheet beds esker networks can extend for hundreds of kilometres, radiating out from the former ice sheet dispersal centres, helping us to reconstruct the meltwater drainage pathways in and beneath the ice (Figure 37). They represent the partially choked ice-walled tunnels that were excavated within and beneath the ice during overall deglaciation. Although single ridges can appear dominant in the landscape, closer inspection often reveals complex networks or braided ridges, where former

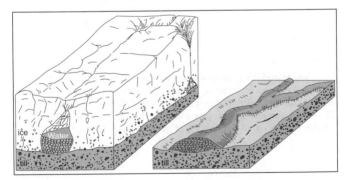

37. **An explanatory sketch to depict the formation of an esker.**

tunnels have migrated and individual, dominant 'trunk' eskers are linked to tributary eskers feeding into them and distributary eskers branch away from them. Long esker networks, potentially stretching hundreds of kilometres, are not deposited at the same time but rather as segments representing the ablation zone of the glacier or ice sheet as it recedes; this is instructive as it shows that the locations of the axes of the main drainage tunnels remain relatively stable over the life cycle of the ice sheet.

Ice-marginal moraines

Throughout the first half of the 20th century glacial geologists routinely referred to two types of moraine as 'end moraine' and 'ground moraine', the former being a landform marking a glacier margin and the latter, introduced by Louis Agassiz, referring to spreads of glacial deposits (tills) lacking any ice-marginal alignment. Therein lay an over-prolonged and misguided conflation of the terms till and moraine, for moraines are very often not composed of till; moreover, the term ground moraine is a classic redundancy, for moraines can never reside in the air!

A range of processes operate at glacier and ice sheet margins and these have been employed to define moraine types genetically. The simplest view of an ice-marginal moraine is to regard it as the end product of a range of debris transfer processes operating in a glacier or ice sheet that deliver material to the snout, akin to a conveyor belt, so that more prolonged periods of ice margin stillstand give rise to larger moraines. The conveyor belt analogy was first coined by the prolific American glacial geologist Warren Upham in the late 19th century, who envisaged englacial debris bands transferring the material to the snout. We now understand that the inexorable deformation of subglacial till towards a glacier snout will also feed the construction of a marginal moraine and that all materials in the path of an active glacier will be subject to pushing or bulldozing by the ice.

Albrecht Penck and Eduard Brückner's *glacial series* (Box 3) depicted the glacial conveyor belt concept in the form of the erosional source basin, the drumlinized glacier bed, and the terminal moraine, but the specific mechanics of moraine production have been only gradually elucidated over two centuries of observation and experiment. In the simplest of terms, after the delivery of sediment to the ice margin by subglacial deformation and/or the melt-out of debris-rich ice, the mechanics can be described as pushing or bulldozing and dumping, often accompanied in poorly drained locations by squeezing. The perceptive glacial geologist Thomas C. Chamberlin was foremost not only in the recognition of the vast end moraines of the Laurentide Ice Sheet in the United States but also in deriving theories on their mode of construction. Chamberlin wrote of the Laurentide end moraines:

> Some of these complex terminal moraines are majestic in dimensions and indescribably intricate in structure, forming great irregular thickened zones of hummocky drift from three to five or even ten or fifteen miles in breadth, and from 100 to 500 ft in depth.

His explanation of their construction involved three process-form relationships, including dump moraine, lodge moraine/sub-marginal moraine, and push moraine. So some eighty years before glacial scientists monitored glacier beds for signs of sediment deformation, T. C. Chamberlin proposed that sub-marginal till deformation had an important role to play in moraine construction.

Pushing or bulldozing is accomplished through the process of glacitectonic deformation by glacier snouts (Box 12) and is responsible for the construction of some impressive moraine types called *composite ridges* and *hill-hole pairs* (Figure 38). Such features are the products of thrusting, folding, and compressive stacking of any deposits or even bedrock that lay in the path of the advancing ice. Where they become overridden by the ice margin that constructed them, composite ridges and hill-hole pairs can become glacially smoothed and capped by a

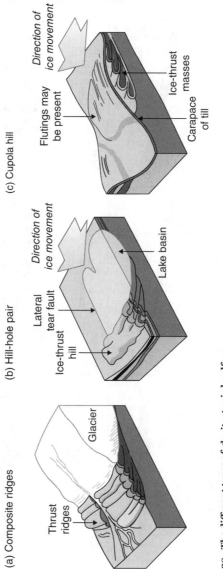

(a) Composite ridges

Thrust ridges

Glacier

(b) Hill-hole pair

Direction of ice movement

Lateral tear fault

Ice-thrust hill

Lake basin

(c) Cupola hill

Direction of ice movement

Flutings may be present

Ice-thrust masses

Carapace of till

38. The different types of glacitectonic landform.

Glaciation

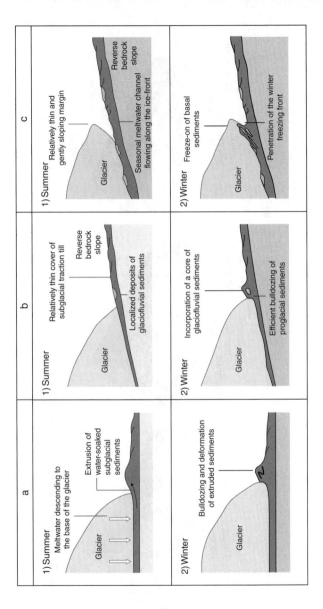

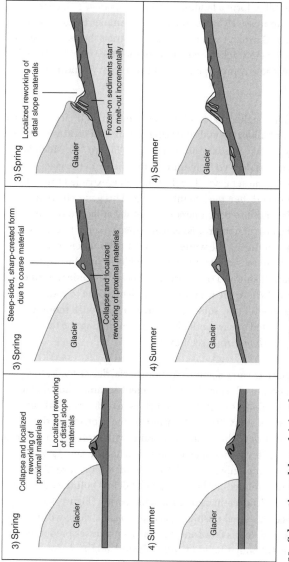

39. Schematic models explaining the processes of minor push moraine formation: (a) efficient bulldozing of extruded subglacial sediments; (b) efficient bulldozing of pre-existing proglacial sediments; (c) emplacement of sediment slabs through subglacial freeze-on.

subglacial till to form a landform called a *cupola hill*. Such features are often impressively large, such as the Dammer Berge Moraine in northern Germany, where six nappes (thrust sheets) each up to 50 m thick were stacked one on top of the other over an area of 12 km^2 by the advancing margin of the former Scandinavian Ice Sheet.

At generally smaller scales (often less than 5 m high), ice-marginal pushing or bulldozing and squeezing of sediment creates features generally known as *minor push moraines* (Figure 39). Much of this material is derived from immediately beneath the glacier snout where the subglacially deforming till is extruded towards the margin and then pushed into a ridge. This generally requires the glacier snout to be temperate or melting at its bed but the moraine construction process can also involve some freezing of the sediment during the winter; in colder environments frozen material is more widespread and hence blocks may be displaced in a similar fashion to that observed in glacitectonic moraine formation. Where the ice margin is indented due to radial crevasses, the till may squeeze upwards to form push moraines with crenulated or sawtooth planforms, prompting the term *sawtooth push moraine* (see Box 2).

Where glaciers carry significant supraglacial debris loads, their margins become the locations of sediment accumulation due to various remobilization processes such as mass flowage, fall, or fluvial transport involved in transferring the debris from the melting ice front. This creates a moraine with probably the longest pedigree, as it was identified by both de Saussure and Agassiz, now more commonly termed *latero-frontal dump moraines*. Their size is related to the supraglacial debris volume and the length of the ice-marginal stillstand period (Figure 40). Where ice-marginal pushing does not create a ridge within such deposits, and debris flows and glacifluvial processes predominate, the sediment bodies build up *latero-frontal fans and ramps*, which consist of coalescent debris fans descending from the

40. Latero-frontal moraines over 100 m high around the snout of Kvíárjökull, Iceland.

glacier snout. Once abandoned by the receding ice, the inner slopes of these fans and ramps are very steep, because they effectively acted as the former ice-contact face, thereby creating a landform with a pronounced asymmetric cross profile.

Supraglacial associations

Sediment–landform associations that originate englacially or supraglacially and then evolve as they are progressively lowered on to the substratum by glacier melting include medial moraines, supraglacial hummocky moraine and controlled moraine, kame and kettle topography, ice-walled lake plains, kame terraces, and pitted sandar. They have much in common, such as being the products of repeated topographic inversion and/or glacier karst development on debris-mantled glaciers, characteristics often used to refer to them collectively as *ice-stagnation topography* (Figure 41). Commonly they are difficult to tell apart because

(a) STAGE 1
Ice stagnation and the formation of supra,
en, and sub glacial channel networks

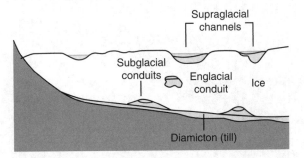

STAGE 2
Development of a complex glacier-karst

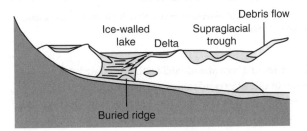

STAGE 3
Topographic inversion

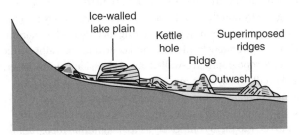

41. a: The development of an ice-contact assemblage of glacifluvial
landforms showing the gradual enlargement of a glacier karst
system during glacier downwasting.

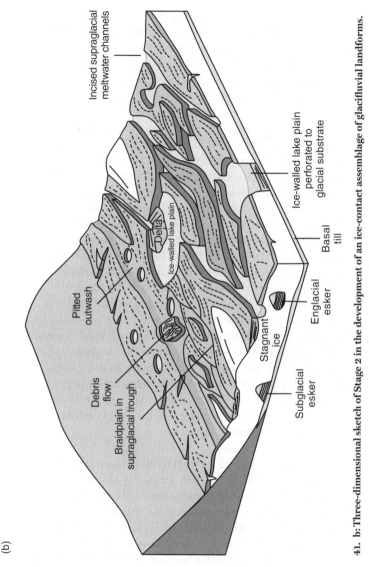

Incised supraglacial meltwater channels

Ice-walled lake plain perforated to glacial substrate

Basal till

Englacial esker

Stagnant ice

Subglacial esker

Pitted outwash

Debris flow

Braidplain in supraglacial trough

Delta

Ice-walled lake plain

(b)

41. b: Three-dimensional sketch of Stage 2 in the development of an ice-contact assemblage of glacifluvial landforms.

they all result from processes that operate in close proximity to one another, such as mass movement, glacifluvial, and glacilacustrine processes, and so hybrid sediment–landform associations are commonplace. It is easier however to describe and explain them separately but to bear in mind that their process-form regimes overlap and some will develop into one another over time.

Medial moraines are prominent on the surfaces of glaciers where ice flow units coalesce or debris repeatedly falls on to the glacier surface in the same area of the accumulation zone (Figure 2), but their preservation potential is low due to their low debris content. After deglaciation, where visible they can be traced as diffuse linear spreads of coarse, often bouldery debris. Supraglacial hummocky moraine and controlled moraine is created by the concentration of debris as it melts out from a debris-charged glacier and is subject to significant amounts of reworking and disturbance before it finally comes to rest on the substrate. This means that the moraine is usually chaotic in form and characterized internally by complex and disturbed sequences of poorly stratified sediments and mass flow diamictons. A case has been made that pronounced transverse linear elements can be preserved and that this variant of hummocky moraine should be called controlled moraine, because the morphology was controlled by the former pattern of debris concentrations in the parent ice, but such features appear to be visible only when ice is still present and hence forming what has traditionally been called ice-cored moraine (see Box 9).

The term kame has a long and complicated history. Derived from the Scottish word 'kaim' it literally means crooked and winding or steep-sided mound and hence its historical conflation with the term esker has been understandable. Nevertheless, kames and eskers are very closely linked genetically and often difficult to differentiate in the landscape where eskers are discontinuous. This complex interrelationship was first addressed with substance by Richard Foster Flint in a series of papers in the 1920s and 1930s, from which derive the ice-contact glacifluvial landform diagrams of many modern general physical geography textbooks.

A kame is essentially any concentration of glacifluvial sediment deposited in contact with melting glacier ice, and they were first observed evolving along modern ice margins by Israel C. Russell during his expedition to the Malaspina Glacier in Alaska in 1890. Kame and kettle topography comprises tracts of mounds and ridges (kames) and intervening hollows (kettles or kettle holes) (Figure 42). The term kettle derives from the fanciful legend that a large hole, especially when containing water, must have served as the devil's water heater! They actually represent areas of subsidence caused by the melt of buried ice, and kame and kettle

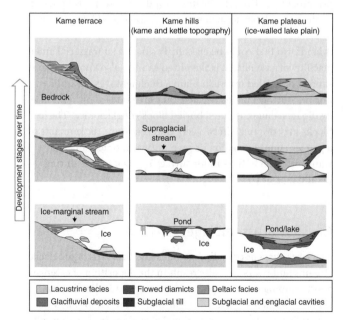

42. The sequential development of the various ice-contact glacifluvial landforms.

topography evolves wherever large quantities of debris are reworked in supraglacial and englacial drainage systems (glacier karst) during the final stages of glacier wastage (Figure 42). The concentration of glacial meltwater and sediment at the edges of valley glaciers will give rise to *kame terraces* as a result of glacifluvial deposition along the downwasting glacier margin.

The ponding of substantial volumes of meltwater on glacier surfaces can result in the production of supraglacial lakes, the deposits of which would be difficult to identify after glacier melt-out. However, a well-developed glacier karst system may create deep perforations in a glacier that can fill with water and ultimately thick sequences of lake sediments. Once the ice walls are removed from such concentrations of sediment they can be left as prominent positive relief features on the landscape, best described as steep-sided, flat-topped mounds or *ice-walled lake plains*.

A landform that could be classified as both a supraglacial and a glacifluvial assemblage is the pitted sandur, also known as kettled sandur or kettled outwash plains. Such features are unlike normal outwash plains because they are cratered by hollows left by the melt-out of isolated buried blocks of glacier ice. The ice blocks may originate in two very different ways: as remnants of a glacier snout buried by its own outwash, or as icebergs transported on to the sandur surface by flood waters during jökulhlaups or GLOFs (Figure 43).

Glacifluvial assemblages

The large suspended sediment loads and bedloads of proglacial streams and rivers typically result in the deposition of extensive, gently sloping outwash plains known by the Icelandic term sandar (singular sandur). As indicated by this adoption of the Icelandic term by the international glacial research community,

43. Aerial view of spectacular iceberg melt-out pits on the surface of the outwash fan created during the 1996 jökulhlaup at Skeiðararjökull, Iceland, with people on the ground for scale.

the first intensive studies of the nature and dynamics of proglacial outwash were undertaken in south-east Iceland, specifically on the Hoffellsandur. The earliest of such studies appears to have been by Norwegian geologist Amund Helland (1846–1918), who in 1882 made the first measurements of glacifluvial sediment transport. This was followed in the late 1930s by the influential Icelandic glacier scientist Sigurður Þórarinsson (1912–83), as part of the Swedish–Icelandic Expedition, and in the early 1950s, by members of the Uppsala University Expedition, under Swedish geographer Filip Hjulström (1902–1982). A decade later, fellow Swede Arne Krigström defined 'valley sandur' and 'plain sandur', the former being valley-confined outwash and the latter being an outwash fan; in North America, T. C. Chamberlin had previously created the term 'valley train' for valley sandur in 1883, a term that has persisted in the North American literature.

The most intensive studies on the pattern of landforms and sediments of sandar have come from Canadian sedimentologist Andrew Miall and British geographer Judith Maizels based upon their scrutiny of modern outwash systems over the period 1970–90. Using sedimentary sequences found in proglacial outwash located at various distances from glacier snouts, Miall has compiled a model of the typical pattern of downstream-fining of glacifluvial deposits that should appear in a single sandur (Figure 44). Downstream-fining and decreasing slope angle with distance were characteristics of outwash fans first noted by Danish geologist Johan G. Forchhammer (1794–1865) in 1847. Miall's model identifies the specific changes in sediment types or 'facies' that are associated with these characteristics and effectively translates a century of field and laboratory based observations on fluvial processes into a framework of predictable sedimentological signatures (Figure 44).

The term 'facies' is used by sedimentologists to describe a sediment body or package whose characteristics are specific to a particular depositional process. It is derived from the Latin term 'facia' relating to how something appears and can be recognized and is applied to a sediment in order to aid its genetic classification. The term 'lithofacies' relates to a sediment when it is defined descriptively and non-genetically (e.g. coarse gravel lithofacies, which could also be classified according to a numbering scheme such as LF 1 etc.). Facies is then employed once a genesis has been interpreted (e.g. fluvial facies, glacigenic facies, etc.).

Glacier-proximal deposits are dominated by poorly sorted, coarse-grained gravels arranged in braided channels and these become finer in a downstream direction, forming numerous sand-filled, shallow distributaries. So sandur sedimentation

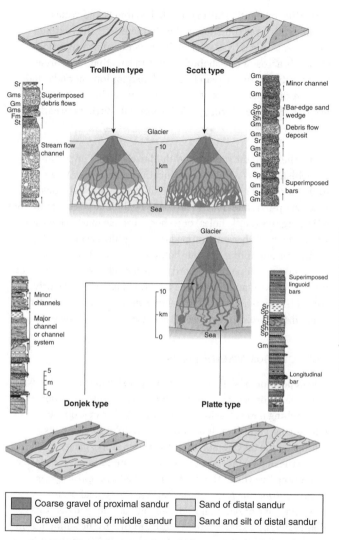

44. Andrew Miall's model from the late 1970s showing typical sandar characteristics based on facies changes. This shows downstream-fining based upon the facies models of four rivers, including the Trollheim, Scott, Donjek, and Platte rivers.

occurs in constantly shifting braided channels and varies significantly over time due to the characteristic fluctuating discharges driven by glacier melting trends. The migration of channels across a fan-shaped sandur means that large areas of the outwash surface can be abandoned for long periods, and even in the active areas of the sandur many channels are only occupied by water during peak flows or only during the early melt season.

Using the Icelandic sandar, Judith Maizels added important embellishments to Miall's classification scheme relating to the impact of glacier floods or jökulhlaups. Most obviously, jökulhlaup-fed sandar are identifiable by their large concentrations of iceberg pits (kettle holes) or iceberg obstacle marks (Figure 43), where the bergs became trapped in the rapidly accumulating sediments and initiated scour holes on the river bed. But they also contain sedimentary records of abrupt changes in facies from normal braided stream gravels to giant slugs of very poorly sorted boulder-rich and matrix-supported gravels, often referred to as 'dirty gravels' and indicative of debris slurry sedimentation.

Subaqueous assemblages

The deposition of sediment by meltwater streams below the water line in lakes or oceans leads to their *progradation* or advance by avalanching and mass flowage into deep water in sequences of clinoforms, similar to those observed in delta foreset beds. Unlike deltas they do not enter the water at its surface and hence the depo-centres or centres of accumulation at the glacier grounding line (floatation point) are called *subaqueous fans* or *grounding line fans*. These features are typical of the proximal zone of sedimentation, and are fan-shaped deposits dominated by coarse-grained sediment due to the sudden drop in stream velocity (Figure 24). As they are fed by meltwater emerging from subglacial tunnels, subaqueous fans may occur at various points along esker ridges to form *esker beads*. Along a stationary ice front the fans may also coalesce to form a particular type of push moraine called a *de Geer moraine* (see Box 10). Where glacially fed meltwater progrades sediment into a lake or the sea it creates a delta, features that form either by the growth of subaqueous fans to create *ice-contact deltas* or at the end of proglacial meltwater streams to create *glacier-fed deltas* (Figure 24).

Chapter 8
Glaciers of the past

When Canadian geologist J. B. Tyrrell proposed his multiple centred Laurentide Ice Sheet in 1898, based upon years of hard-won data collection, delivered by long canoe traverses and frugal field camps, he could not have imagined the ease with which we can now view continent-sized swaths of glacial geomorphology using satellite-based imagery. Yet his reconstruction endures. What does this tell us? It is certainly testament to the invaluable nature of diligent and systematic field-based empirical data collection but it also requires a talent to see the bigger picture.

For more recent generations of glacial geomorphologists this message was reiterated and a challenge thrown down by the American glacial researchers George Denton and Terence Hughes in 1981 when they published *The Last Great Ice Sheets*. At this time field workers were slugging it out repeatedly in highly critical debates about how extensive the last ice sheets had been, to the extent where the glacial research community could be subdivided into the 'maximalists' and the 'minimalists'. Indeed, the debate continued after 1981 and somewhat mischievously Hughes implied that maybe the minimalists were not seeing all the diagnostic signs:

> opposing views on this question will never be reconciled because
> they are fixed in our minds not by field evidence but by our own

genes....some of our ancestors...learned to cope with the advancing ice, but...[a timid group] migrated southward and tried to blot the memory of that fearful advance from their minds....after the last major...glaciation, the timid population group migrated back. As a consequence, it is now impossible to distinguish the two population groups. However, genetic recombinations will continue to produce individuals who carry genes that were mostly associated with either the bold or with the timid group. That is why I believe that the debate between maximum and minimum versions of Late Quaternary glaciation will never be resolved. If the above hypothesis is correct, then those who support minimum versions of Late Quaternary ice sheets actually prefer a world in which there are no ice sheets at all, and never were any such things. Unfortunately for them, the evidence for ice sheets, past and present, is all too clear, so they cannot just deny their existence. The next best thing is to deny part of their existence, especially the existence of parts of former ice sheets. It is much easier to deny the existence of an ice sheet that is now gone, than one that is still here.

In this light-hearted way the modern era of ice sheet reconstruction was reinvigorated. Clearly Hughes regarded himself, Denton, and their colleagues as the bold, maximalist 'descendants' and compiled theoretical models of pervasive former ice sheets whose dynamics were compatible with what we knew of contemporary ice sheet systems, especially the functioning of ice streams and multiple dispersal centres. So *The Last Great Ice Sheets* contained not quite outrageous geological hypotheses as defined by W. M. Davis but certainly hypotheses that seriously challenged the field working glacial geomorphologist.

In fairness to the minimalists, or indeed any glacial geomorphologist at this juncture, their reconstructions were based upon the level of sophistication of their process-form-based framework, the details of which were still evolving. But to reconstruct the former extent and dynamics of ice sheets and glaciers requires a knowledge of process-form relationships that goes beyond individual landform types. Instead, glacial geomorphologists need to analyse large areas of glaciated terrain in a more holistic way, as J. B. Tyrrell had demonstrated, combining the whole range of glacial landforms and sediments to reconstruct glacier systems of the past, a subject now known as *palaeoglaciology*.

Once the first aerial photographs were taken of the vast and often uninhabited areas of glaciated terrain in the 1940s more holistic assessments of glacial geomorphology became feasible, a situation that would only improve as more aerial imagery, especially satellite-borne imagery, became available. Also important was the inheritance by palaeoglaciologists of a suitable vehicle of landform analysis that was also first introduced in the 1940s and called the *landsystem*. This is a terrain evaluation procedure wherein the geomorphology and subsurface materials that characterize a landscape are genetically related to the processes involved in their development. A landsystem is an area of common terrain attributes, different from those of adjacent areas, in which recurring patterns of topography, soils, and vegetation reflect the underlying geology, past erosional and depositional processes, and climate.

The first glacial landsystems were not actually called landsystem per se but process-form models, the benchmark example being that for the former southern margin of the Laurentide Ice Sheet in North Dakota, compiled by Lee Clayton and Stephen Moran in 1974. Variants of this model or landsystem are recognizable to anyone who has studied physical geography at school, as it was reproduced as a typical lowland glaciation model and compared to the equally widely recognizable mountain glaciation model compiled by W. M. Davis. The first landsystem model compiled from a modern glacier margin was that of R. J. Price in 1969, in which the evolving landforms on the foreland of Breiðamerkurjökull in Iceland were used to classify the landform-sediment assemblages of a temperate glacier snout.

Since the early 1970s a wide range of glacial landsystems have been compiled to represent process-form signatures of particular styles of glaciation. These include styles dictated by topography and proximity to deep water, such as glaciated valley, plateau icefield, volcano-centred, fjord, continental shelf, and glacilacustrine landsystems, and those dictated by glacier dynamics and temperature, such as surging, active temperate, subpolar, polar, ice stream, and GLOF-dominated landsystems (Figure 45). Each landsystem serves as an analogue or exemplar for glacial geomorphologists to use in reconstructing glaciers and ice sheets of the past or palaeoglaciology.

Because the style of glaciation can change temporally as well as spatially, it is important to acknowledge that glacial landsystem signatures will also change. For example, as a landscape is subject to increasingly colder conditions at the start of a glaciation it can develop first plateau icefields, then plateau icefields with outlet valley glaciers and piedmont lobes, and ultimately an ice sheet with ice streams as a result of the coalescence of icefields. Spatially the landsystem imprint can also change as an ice sheet thickens and develops shifting ice dispersal centres and flow patterns. A very clear illustration of this is the overprinting of subglacial bedforms whereby subglacial landforms such as drumlins, flutings, and MSGL can be seen to cross-cut each other. These patterns of overprinting can be mapped out over large areas so that separate families or swarms of bedforms can be assigned to *flowsets*, each flowset recording a prominent ice flow direction and often relating to shifts in ice stream positions.

The record of this flow switching can only ever be partial, because the bedforms of older flow phases will eventually be smudged to varying degrees so that that they will be heavily overprinted and therefore very faint or may disappear altogether from the landform record. The term *glacial landscape palimpsest* has been used to classify this superimposition of the records of former

Glaciation

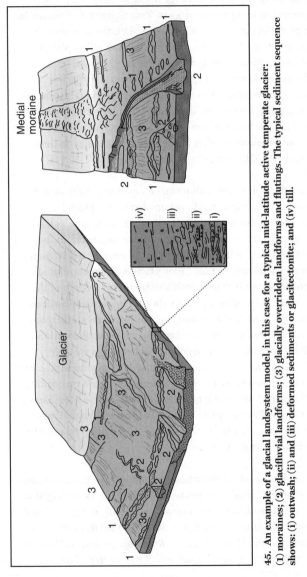

45. An example of a glacial landsystem model, in this case for a typical mid-latitude active temperate glacier: (1) moraines; (2) glacifluvial landforms; (3) glacially overridden landforms and flutings. The typical sediment sequence shows: (i) outwash; (ii) and (iii) deformed sediments or glacitectonite; and (iv) till.

glacier activity, after the term palimpsest to describe a parchment that was reused by early monastic scholars when writing materials were in short supply.

Once mapped out, the footprint of a former glacier can be used to make a palaeoglaciological reconstruction that provides information on not just ice extent but also its surface contours and its palaeo-ELA. Palaeo-ice sheet reconstructions became a reality when the first regional maps of glacial features were compiled. A benchmark example was the compilation of the first *Glacial Map of Canada* in 1959, followed by its update in 1968, by the Geological Survey of Canada, based upon painstaking aerial photograph and field mapping by its officers on a map sheet by map sheet basis after the completion of the aerial photograph coverage for the whole country in the 1950s. Once this was complete it then laid the foundations for the reconstruction of the whole ice sheet and its recession during the last glaciation using a huge database of chronological controls (e.g. radiocarbon dates) on ice-marginal positions. More intensive mapping of glacial landforms from ever-improving aerial and remotely sensed imagery as well as offshore sonar-based swath bathymetry has resulted in ice sheet reconstructions depicting increasingly more complex dynamics (Figure 46).

At the same time, increasingly powerful computers are making it possible to compile sophisticated numerical models that use our knowledge of glaciological processes and ice-core-derived palaeoclimate data to create three-dimensional glacier and ice sheet reconstructions. An excellent example of one such model is that of Alun Hubbard and colleagues of Aberystwyth University, who have created a numerical model of the British-Irish Ice Sheet for the last glaciation. This is very useful because it reveals how glacial landsystem imprints should change in space and time and how a glaciation is dominated not by the most extensive ice sheet cover but by mountain icefield cover, and hence this style of

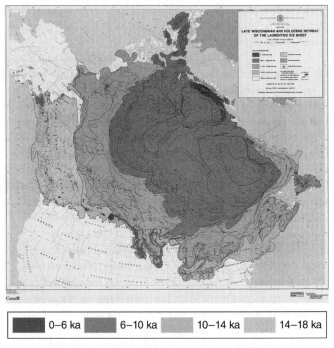

| ■ 0–6 ka | ■ 6–10 ka | 10–14 ka | 14–18 ka |

46. An extract from one of the Dyke and Prest (1987) Laurentide Ice Sheet reconstruction maps, showing the margin recession pattern over time.

glacierization is representative of average glacial conditions. It also shows how shifting ice dispersal centres drive the repeat direction changes of ice streams, thereby verifying glacial landsystem mapping in the field.

Chapter 9
Glaciers, humans, and enduring ice

A knowledge of glaciation is important not only as an academic pursuit but also because it provides us with an understanding of glaciers as earth surface systems in a warming world and of the glacial materials that lie beneath our feet. Crucial in this respect are glacier-related hazards impacting directly on human society and glacial landforms and sediments lying at the surface of some of the most densely populated parts of our planet.

Hazardous glaciers

At 4.30 p.m. on 16 June 1818, the avalanche apron of the snout of the Giétro Glacier in Valais in the Swiss Pennine Alps was breached by the waters of a lake that it had dammed in an adjacent valley. Although attempts had been made to syphon off the water by drilling a tunnel through the apron, it was not enough to stop 18 million m^3 of water catastrophically draining into the lower valleys and killing forty-four people. This scenario was not new to the local population; the same thing had occurred in 1595, when it killed 140 people, and also only one year earlier when it resulted in no fatalities. This is just one of many examples of the glacier-related hazards called GLOFs that impact most often and sometimes most tragically on human society.

GLOFs are well documented in the European Alps, where they are called *débâcles*, in the South American Andes, where they are called *alluviones*, and in the Himalaya and North American Rockies. In each of these regions the increasing potential for, and number of, GLOFs has increased over the last century as the mountain valley glaciers have undergone accelerated recession since the culmination of the Little Ice Age. The nature of the damming has changed over this time, from the damming of valleys by glacier ice to the creation of *moraine-dammed lakes* trapped between the substantial latero-frontal moraines of the Little Ice Age and the downwasting snouts. The breaching of the moraine dam by lake waters can result in catastrophic drainage and can be brought about simply by rising lake levels or by *seiche waves* generated by rock and ice avalanches or even seismic shocks. The resulting GLOF comprises a highly destructive mixture of water, rock, mud, and glacial ice and is augmented by bank erosion along the flood path.

Such a flood in the Cordillera Blanca in Peru in 1941 was of colossal proportions, initiated by an ice avalanche into one moraine-dammed lake and then the escape of water into another moraine-dammed lake down-valley, creating 8 million m^3 of water and debris which killed around 5,000 people in the town of Huaraz, 23 km downstream. Even when lives are not lost, GLOFs can wreak havoc on local communities due to the erosion of cultivated land and destruction of buildings and bridges and important infrastructure such as hydroelectric plants.

A particular type of glacier-induced flood occurs regularly beneath the volcano-centred ice caps of Iceland, where the term *jökulhlaup* has traditionally been used to describe such events. Typical of such events and one of the best documented was that of the morning of 5 November 1996, when a long-overdue jökulhlaup erupted through the snout of Skeiðarárjökull. The volcanic eruption responsible for the flood had taken place between 30 September and 13 October at the Bárðabunga volcanic centre, melting a vast quantity of ice which migrated from the eruption site into subglacial lake Grímsvötn. From there it catastrophically drained, taking only ten hours to reach the glacier margin at a maximum flow rate of 50,000 m^3 per second. Sediment-laden water burst from multiple ice-roofed vents and surface fractures along the entire ice margin and angular blocks of ice in excess of 30 m in diameter were dislodged from the glacier snout. Icebergs liberated from the fractured snout were up to 2,000 m^3 in size and many up to 200 m^3 collided with and destroyed the road bridge on the country's main highway.

Icelanders are accustomed to such events and engineer their infrastructure in such a way as to minimize jökulhlaup impacts and/or they take evasive action when volcanic activity appears to be posing a threat. This is just one aspect of living with glaciers in your neighbourhood and Icelanders and many mountain communities around the world are reliant upon normal glacial meltwater flows for their agriculture and groundwater needs.

It is not just glacier ice and its meltwater but also glacial landforms and sediments that are integral to the everyday lives of people who live in glaciated landscapes, whether they realize it or not. An alarming illustration of this was provided in 1978 when a farmer in Rissa, Norway initiated a catastrophic failure in raised glacimarine deposits by merely excavating a hole for the foundations of a new building. By simply loading the sediments with the spoil from the hole, the farmer initiated mass liquefaction or flowage, and 6 million m^3 of glacimarine clay collapsed over an area of 330,000 m^2, creating a 1.5-km-long slide face and killing one person and consuming thirteen farms, two homes, and a community centre. It also created a 3-m-high flood wave that breached the opposite bank of the lake. Such *quick clays* are often the cause of huge catastrophic failures, because the gradual groundwater removal of salt and hence the elimination of the internal cohesive bonds from the marine sediments after they are lifted above the sea by glacioisostatic rebound makes them susceptible to liquefaction when later loaded.

Other engineering properties of glacial deposits such as the *over-consolidation* and discontinuous stratification of tills are significant when excavating them for construction purposes. Over-consolidation is particularly diagnostic of tills, because it relates to their decrease in volume due to compression by glacial loading. This means that their present-day consolidation state is greater than would be normally expected and so it is crucial that engineers know the exact origins of diamictons on any construction site. Glacitectonic structures and buried pockets of glacifluvial sediments are also important considerations for engineers, because they constitute potential sources of sediment failure and seepage in artificially cliffed glacial deposits.

Glaciation's valuable legacy

The distribution of glacial deposits is important not just for engineering practicalities but also for a range of human activities such as aggregate extraction, groundwater quality, and the planning of potential landfill sites. As the separation of sediment grain sizes for aggregate use is an expensive business, a knowledge of glacifluvial landforms and deposits and hence the locations of sand and gravel is critical to successful aggregate exploration and use. This is exemplified by the clear targeting of esker and kame landforms for quarrying and extraction for the building trade. A stark example is the Carstairs Kames (actually an esker complex), located between Edinburgh and Glasgow, Scotland. The view from the train window of this feature has changed fundamentally in some locations over the last century, as sand and gravel extraction has gradually eliminated many ridges.

Even tills, with their often less useful range of grain sizes, and to a smaller extent glacilacustrine deposits, have been the source of great wealth through the early activities of the brick-making industry, immediately obvious in the siting of former brick pits in areas of clay-rich tills and former glacial lake plains. The Don Valley Brick Works near Toronto, Canada is the site of thick glacilacustrine clays, which were exploited for brick making in the early 20th century and overlie one of the most important interglacial peat deposits in North America dating to 120,000 years ago.

The juxtaposition of glacifluvial deposits with less permeable tills, especially in vertical stratigraphic sequences, is less useful for aggregate extraction but significant when considering the location and extent of *aquifers* for groundwater exploitation through water well extraction. Aquifers are sediment bodies that contain and drain large volumes of water because they are more permeable than surrounding materials. In regions where good-quality clean water is at a premium, the occurrence of extensive sand and gravel beds between tills becomes crucial to agriculture and urban water supplies and in some settings may even contain the residual meltwaters from the last ice sheets.

The importance of such aquifers is well illustrated on the prairies of west-central North America, where the Laurentide Ice Sheet has repeatedly invaded a flat-lying terrain and plastered tills over interglacial river gravels and glacial meltwater deposits. In such landscapes, an understanding of the complexities of glacial stratigraphy is also crucial to the planning and execution of landfill, particularly in relation to the potential for groundwater contamination; it is potentially very hazardous to assume that a lowland glaciated landscape is simply underlain by thick sequences of monotonous, consolidated, clay-rich till, which a landfill company would treasure as the perfect seal to prevent the downward penetration of poisonous leachate.

Even where glacial deposits blanket the underlying bedrock surface and hence hide it from our exploitative eyes, the riches of mineral veins and ore bodies within that bedrock can be pinpointed with remarkable accuracy if our knowledge of glacial processes is applied to the problem. This has been demonstrated many times in a most lucrative way on the Canadian and Scandinavian Shields through a procedure called 'till prospecting'. This has often been termed 'following glacial breadcrumbs', for it refers to the systematic sampling of tills over large areas to see if they have incorporated valuable mineral deposits (e.g. gold, uranium, silver, nickel, diamonds) and if so whether a knowledge of glacial flow direction can be used to locate the origins of such treasures.

Some of the most valuable mineral concentrations on the planet have been located by glacial geomorphologists, sampling tills over large areas and reconstructing former ice flow patterns from ice sheet footprints. In our older glacial record, it is the former sands and gravels of ancient glaciations that have become a target for hydrocarbon reservoir exploitation. Examples include Ordovician grounding line fans and glacifluvial outwash in North Africa, Cryogenian tunnel valleys in the São Francisco River basin in Brazil, and Permo-Carboniferous glacial valley infills in Bolivia.

Our glacial planet: does ice have a future?

It does not really matter where you are on the planet, glaciers have probably influenced it in some way, if not fundamentally, at some point in the Earth's long history. At the longest of timescales the outcrops of 2-billion-year-old glacial deposits allow us to stare into deep time when an iceberg dropped a pebble for it to drop into the bottom muds in a world devoid of anything other than single-celled life forms or bacteria. The last glaciation occurred on our watch and our ancestors had to adapt to the increasing frigidity of high- to mid-latitude regions. But we have more recently entered into what many regard as a new geological epoch, one that has been dubbed the 'Anthropocene' because it is the first time in the Earth's long geological history that humans have influenced environmental and climate conditions.

Opinions differ on how far back in time we should place the lower boundary of this new epoch, with the longest timescale being 12,000 years ago, at the dawn of agriculture. On one thing we can be certain of, since the Little Ice Age glacier ice has been on the retreat at an increasingly rapid rate and this loss is being measured also in sea level rise as meltwater is released to our oceans; total glacier melt created a sea level rise of more than 1 mm per year over the period 1993–2003.

But is Earth entering a new phase of ice free conditions, the like of which it has not endured for more than 35 million years? At present rates of retreat, it is likely that in 2100 many mountain glaciers may have disappeared, and mid-latitude ice caps such as those in the European Alps, Iceland, and New Zealand could be mere remnants of their present incarnations, although it is very difficult to say whether or not ice would disappear altogether from large parts of the planet's presently glacierized areas.

The Antarctic Ice Sheet has endured for the longest of all our present ice bodies, at least 700,000 years according to the ice core records but likely for much longer, at least 30 million years. Our oldest surviving ice appears to be that buried beneath a thick supraglacial debris cover in East Antarctica and could be a staggering 8 million years old. Clearly it would take very significant climate warming to remove such glacier ice remnants and hence our planet is likely to be in an icehouse state for some time to come.

References

A vast literature has accumulated on the subjects of glaciology and glacial geomorphology and hence any list of references used in the compilation of this book will always be very selective. It is recommended therefore that those who would like to delve more deeply into the academic side of this fascinating area of study should start with the books listed in the further reading section and follow the threads of the literature cited therein—enjoy the journey!

The references constitute the author's personal selection of academic papers which, although they form only part of the substantial collection used in researching this book, serve to provide a manageable selection of examples of the concepts introduced here. They also represent a historical journey through the development of our understanding of glaciation. For a more thorough overview of the development of the glacial theory, the *Very Short Introduction* on *The Ice Age* by Jamie Woodward is highly recommended and indeed forms an appropriate companion volume to this one.

Alt, D. 2001. *Glacial Lake Missoula and its Humongous Floods*. Mountain Press, Montana.

Benn, D. I. 1992. Scottish landform examples—5: The Achnasheen terraces. *Scottish Geographical Magazine* 108, 128–31.

Boulton, G. S. 1986. A paradigm shift in glaciology? *Nature* 322, 18.

Boulton, G. S. 1987. A theory of drumlin formation by subglacial sediment deformation. In Menzies, J. and Rose, J. (eds), *Drumlin Symposium*. Balkema, Rotterdam, pp. 25–80.

Boulton, G. S. and Hindmarsh, R. C. A. 1987. Sediment deformation beneath glaciers: rheology and sedimentological consequences. *Journal of Geophysical Research* 92, 9059–82.

Chamberlin, T. C. 1894. The glacial phenomena of North America. In Geikie, J., *The Great Ice Age*. Stanford, London, pp. 724–75.

Church, M. and Ryder, J. M. 1972. Paraglacial sedimentation: a consideration of fluvial processes conditioned by glaciation. *Geological Society of America Bulletin* 83, 3059–67.

Clarke, G. K. C. 1987. A short history of scientific investigations on glaciers. *Journal of Glaciology*. Special Issue, 4–24.

Cunningham, F. 1990. *James David Forbes—Pioneer Scottish Glaciologist*. Scottish Academic Press, Edinburgh.

Ehlers, J., Gibbard, P. L., and Hughes, P. D. (eds) 2016. *Quaternary Glaciations: Extent and Chronology—A Closer Look*. Elsevier.

Embleton, C. and King, C. A. M. 1976. *Glacial Geomorphology*. Arnold, London.

Evans, D. J. A. (ed.) 2004. *Geomorphology: Critical Concepts in Geography—Volume IV, Glacial Geomorphology*. Routledge, London.

Evans, D. J. A. 2008. Glacial depositional processes and forms. In Burt, T. P., Chorley, R. J., Brunsden, D., Cox, N. J., and Goudie, A. S. (eds), *The History of the Study of Landforms: Volume 4—Quaternary and Recent Processes and Forms (1890–1965) and the Mid-Century Revolutions*. Geological Society, London, pp. 495–619.

Evans, D. J. A. and Benn, D. I. (eds) 2004. *A Practical Guide to the Study of Glacial Sediments*. Arnold, London.

Evans, I. S. 2008. Glacial erosional processes and forms: mountain glaciation and glacier geography. In Burt, T. P., Chorley, R. J., Brunsden, D., Cox, N. J., and Goudie, A. S. (eds), *The History of the Study of Landforms: Volume 4—Quaternary and Recent Processes and Forms (1890–1965) and the Mid-Century Revolutions*. Geological Society, London, pp. 413–94.

Fairchild, H. L. 1907. Drumlins of central western New York. *New York State Museum Bulletin* 111, Albany, New York.

Flint, R. F. 1971. *Glacial and Quaternary Geology*. Wiley, New York.

Forbes, J. D. 1859. *Occasional Papers on the Theory of Glaciers*. Edinburgh.

Geikie, J. 1894. *The Great Ice Age and its Relationship to the Antiquity of Man*, 3rd edition. Stanford, London.

Harland, W. B. 2007. Origins and assessment of snowball Earth hypothesis. *Geological Magazine* 144, 633–42.

Hubbard, A., Sugden, D., Dugmore, A., Norddahl, H., and Pétursson, H. G. 2006. A modelling insight into the Icelandic Last Glacial Maximum ice sheet. *Quaternary Science Reviews* 25, 2283–96.

Hughes, A. L. C., Gyllencreutz, R., Lohne, Ø. S., Mangerud, J., and Svendsen, J. I. 2016. The last Eurasian ice sheets—a chronological database and time-slice reconstruction, DATED-1. *Boreas* 45, 1–45.

Lewis, W. V. (ed.) 1960. *Norwegian Cirque Glaciers*. Royal Geographical Society Research Series 4. Royal Geographical Society, London.

Nye, J. F. 1948. The flow of glaciers. *Nature* 161, 819–21.

Palsson, S. 1795. (English translation 2004). *Draft of a Physical, Geographical and Historical Description of Icelandic Ice Mountains on the Basis of a Journey to the Most Prominent of Them in 1792–1794*. The Icelandic Literary Society, Reykjavik.

Price, R. J. 1969. Moraines, sandar, kames and eskers near Breiðamerkurjökull, Iceland. *Transactions of the Institute of British Geographers* 46, 17–43.

Price, R. J. 1973. *Glacial and Fluvioglacial Landforms*. Longman, London.

Sugden, D. E. and John, B. S. 1976. *Glaciers and Landscape*. Arnold, London.

Tyrrell, J. W. 1897. *Across the Sub-Arctics of Canada*. William Briggs, Toronto.

Woodward, J. C. 2014. *The Ice Age: A Very Short Introduction*. Oxford University Press, Oxford.

Further reading

Alley, R. B. 2000. *The Two-Mile Time Machine*. Princeton University Press, Princeton.

Andersen, B. G. and Borns, H. W. 1994. *The Ice Age World*. Scandinavian University Press, Oslo.

Benn, D. I. and Evans, D. J. A. 2010. *Glaciers and Glaciation*. 2nd edition. Hodder Education, London.

Bennett, M. R. and Glasser, N. F. 2009. *Glacial Geology: Ice Sheets and Landforms*. 2nd edition. Wiley Blackwell, Oxford.

Ehlers, J., Hughes, P., and Gibbard, P. L. 2016. *The Ice Age*. Wiley Blackwell, Oxford.

Evans, D. J. A. (ed.) 2003. *Glacial Landsystems*. Arnold, London.

Evans, D. J. A. 2016. *Vatnajokull National Park (South Region)—Guide to a Glacial Landscape Legacy*. Vatnajokull National Park, Reykjavik.

Giardino, R. and Harbor, J. (eds) 2013. *Treatise on Geomorphology—Volume 8: Glacial and Periglacial Geomorphology*. Academic Press, San Diego.

Griffiths, J. S. and Martin, C. J. (eds) 2017. *Engineering Geology and Geomorphology of Glaciated and Periglaciated Terrains*. Geological Society Engineering Geology Special Publication 28, London.

Hambrey, M. J. and Alean, J. 2004. *Glaciers*, 2nd edition. Cambridge University Press, Cambridge.

Hambrey, M. J. and Alean, J. C. 2017. *Colour Atlas of Glacial Phenomena*. CRC Press, Boca Raton.

Knight, P. G. 1999. *Glaciers*. Stanley Thornes, Cheltenham.

Knight, P. G. (ed.) 2006. *Glacier Science and Environmental Change*. Blackwell, Oxford.

Nesje, A. and Dahl, S. V. 2000. *Glaciers and Environmental Change*. Arnold, London.

Post, A. and LaChapelle, E. R. 1971. *Glacier Ice* (revised in 2000). University of Washington Press, Seattle.

Whiteman, C. A. 2011. *Cold Region Hazards and Risks*. Wiley-Blackwell, Oxford.

Additionally, a particularly useful compendium of up-to-date reviews of a full range of topics related to the Quaternary Period and its glaciations is the *Encyclopedia of Quaternary Science*, edited by S. A. Elias (2013) and published by Elsevier.

Some useful websites

Four very watchable short lectures on introducing the basic principles and processes of glaciers called *Crash Course Cryosphere 1–4* are available on YouTube at: <https://www.youtube.com/watch?v=zU46kdTtlUk>.

All the details of the Storglaciaren mass balance record can be viewed on the website of the Bolin Centre for Climate Research at Stockholm University: <http://bolin.su.se/data/tarfala/storglaciaren.php>.

The World Glacier Monitoring Service (WGMS) can be accessed here: <http://wgms.ch/>.

A fascinating fifty-minute lecture by John Nye entitled *Glaciology 65 years ago* can be viewed at: <https://youtu.be/5w38d4GL2O4>.

The BBC TV series *Men of Rock*, celebrating the contributions of Scottish geologists to the Earth Sciences, is available on YouTube, with the third programme on *The Big Freeze* being most relevant to this book: <https://www.youtube.com/watch?v=RIjFP6FJTyc>.

An animation of the numerical model of the British-Irish Ice Sheet compiled by Alun Hubbard and his co-workers is well worth watching and is available to download and view at: <https://www.aber.ac.uk/en/dges/research/centre-glaciology/research-intro/biis/>.

A fascinating film of the quick clay landslide at Rissa, captured by amateur footage at the time, is available on YouTube. This includes a laboratory-based explanation of how quick clays create landslides: <https://www.youtube.com/watch?v=3q-qfNlEP4A>.

Index

Glaciation

DESERTS
A Very Short Introduction
Nick Middleton

Deserts make up a third of the planet's land surface, but if you picture a desert, what comes to mind? A wasteland? A drought? A place devoid of all life forms? Deserts are remarkable places. Typified by drought and extremes of temperature, they can be harsh and hostile; but many deserts are also spectacularly beautiful, and on occasion teem with life. Nick Middleton explores how each desert is unique: through fantastic life forms, extraordinary scenery, and ingenious human adaptations. He demonstrates a desert's immense natural beauty, its rich biodiversity, and uncovers a long history of successful human occupation. This *Very Short Introduction* tells you everything you ever wanted to know about these extraordinary places and captures their importance in the working of our planet.

www.oup.com/vsi

FORENSIC SCIENCE
A Very Short Introduction
Jim Fraser

In this Very Short Introduction, Jim Fraser introduces the concept of forensic science and explains how it is used in the investigation of crime. He begins at the crime scene itself, explaining the principles and processes of crime scene management. He explores how forensic scientists work; from the reconstruction of events to laboratory examinations. He considers the techniques they use, such as fingerprinting, and goes on to highlight the immense impact DNA profiling has had. Providing examples from forensic science cases in the UK, US, and other countries, he considers the techniques and challenges faced around the world.

> An admirable alternative to the 'CSI' science fiction juggernaut...Fascinating.
>
> **William Darragh, Fortean Times**

www.oup.com/vsi

GALAXIES
A Very Short Introduction
John Gribbin

Galaxies are the building blocks of the Universe: standing like islands in space, each is made up of many hundreds of millions of stars in which the chemical elements are made, around which planets form, and where on at least one of those planets intelligent life has emerged. In this *Very Short Introduction*, renowned science writer John Gribbin describes the extraordinary things that astronomers are learning about galaxies, and explains how this can shed light on the origins and structure of the Universe.

GEOGRAPHY
A Very Short Introduction
John A. Matthews & David T. Herbert

Modern Geography has come a long way from its historical roots in exploring foreign lands, and simply mapping and naming the regions of the world. Spanning both physical and human Geography, the discipline today is unique as a subject which can bridge the divide between the sciences and the humanities, and between the environment and our society. Using wide-ranging examples from global warming and oil, to urbanization and ethnicity, this *Very Short Introduction* paints a broad picture of the current state of Geography, its subject matter, concepts and methods, and its strengths and controversies. The book's conclusion is no less than a manifesto for Geography' future.

'Matthews and Herbert's book is written- as befits the VSI series- in an accessible prose style and is peppered with attractive and understandable images, graphs and tables.'

Geographical.

LANDSCAPES AND GEOMORPHOLOGY
A Very Short Introduction

Andrew Goudie & Heather Viles

Landscapes are all around us, but most of us know very little about how they have developed, what goes on in them, and how they react to changing climates, tectonics and human activities. Examining what landscape is, and how we use a range of ideas and techniques to study it, Andrew Goudie and Heather Viles demonstrate how geomorphologists have built on classic methods pioneered by some great 19th century scientists to examine our Earth. Using examples from around the world, including New Zealand, the Tibetan Plateau, and the deserts of the Middle East, they examine some of the key controls on landscape today such as tectonics and climate, as well as humans and the living world.

www.oup.com/vsi

SLEEP
A Very Short Introduction

Russell G. Foster & Steven W. Lockley

Why do we need sleep? What happens when we don't get enough? From the biology and psychology of sleep and the history of sleep in science, art, and literature; to the impact of a 24/7 society and the role of society in causing sleep disruption, this *Very Short Introduction* addresses the biological and psychological aspects of sleep, providing a basic understanding of what sleep is and how it is measured, looking at sleep through the human lifespan and the causes and consequences of major sleep disorders. Russell G. Foster and Steven W. Lockley go on to consider the impact of modern society, examining the relationship between sleep and work hours, and the impact of our modern lifestyle.

www.oup.com/vsi

WITCHCRAFT
A Very Short Introduction
Malcolm Gaskill

Witchcraft is a subject that fascinates us all, and everyone knows what a witch is - or do they? From childhood most of us develop a sense of the mysterious, malign person, usually an old woman. Historically, too, we recognize witch-hunting as a feature of pre-modern societies. But why do witches still feature so heavily in our cultures and consciousness? From Halloween to superstitions, and literary references such as Faust and even Harry Potter, witches still feature heavily in our society. In this Very Short Introduction Malcolm Gaskill challenges all of this, and argues that what we think we know is, in fact, wrong.

'Each chapter in this small but perfectly-formed book could be the jumping-off point for a year's stimulating reading. Buy it now.'

Fortean Times